HOW TO RAISE
HORSES

EVERYTHING YOU NEED TO KNOW

BREED GUIDE & SELECTION
PROPER CARE & HEALTHY FEEDING
BUILDING FACILITIES AND FENCING
SHOWING ADVICE

Daniel and Samantha Johnson

Voyageur Press

First published in 2007 by MBI Publishing Company LLC and Voyageur Press, an imprint of MBI Publishing Company, Galtier Plaza, Suite 200, 380 Jackson Street, St. Paul, MN 55101-3885 USA

Voyageur Press titles are also available at discounts in bulk quantity for industrial or sales-promotional use. For details write to Special Sales Manager at MBI Publishing Company, Galtier Plaza, Suite 200, 380 Jackson Street, St. Paul, MN 55101-3885 USA.

To find out more about our books, join us online at www.voyageurpress.com.

About the author:
Daniel Johnson specializes in equine photography but also enjoys photographing many other subjects, such as dogs, farm animals, gardens, and rural life. His images appear in magazines, books, greeting cards, and calendars nationwide. Dan also manages the family-owned horse farm and oversees the breeding, training, and showing of their horses.

Samantha Johnson is a freelance writer and a certified horse show judge. She also works at Fox Hill Farm.

On the cover:
When purchasing a broodmare, you will want to evaluate the quality of her offspring, in addition to the mare's own conformation, breed type, and movement. Ask the owner for photos or video of her previous foals so that you can determine whether she is producing the type and quality that you want. As you can see, this mare has reproduced herself very closely in her foal.

On the back cover:
Bottom: This horse exhibits a slightly interested but reserved expression. The horse is watching something, possibly his owner or another animal who has aroused his curiosity.
Top: Your horse will certainly be waiting for you if he knows he will be fed at a particular time. Horses love to figure out the predictability of their feeding schedule.
Middle: While it may be tempting to purchase an adorable foal for your child, the idea of letting them grow up together is usually misguided. For a child's mount, experience and temperament are more important characteristics to consider.

Library of Congress Cataloging-in-Publication Data

Johnson, Samantha.
  How to raise horses : everything you need to know / text by Samantha Johnson ; photos by Daniel Johnson.
    p. cm.
  ISBN-13: 978-0-7603-2719-7 (softbound)
  ISBN-10: 0-7603-2719-X (softbound)
 1. Horses. I. Johnson, Daniel, 1984- II. Title.
SF285.3.J637 2007
636.1--dc22

                                    2006024857

Editor: Amy Glaser
Designer: Brenda C. Canales

Printed in China

CONTENTS

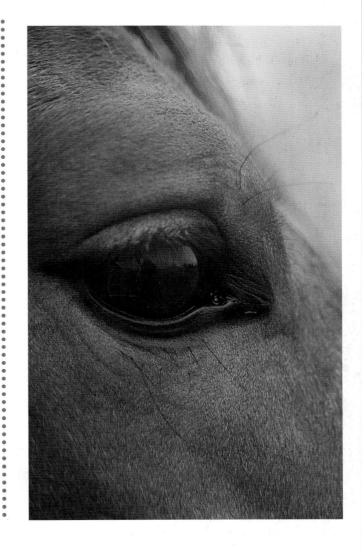

ACKNOWLEDGMENTS

• •

We would like to express our gratitude to the following folks who were of invaluable help during the process of writing this book: Amy Glaser, our editor, for her encouragement, assistance, and suggestions; we appreciate your enthusiasm. Jayme VanHaverbeke Nelson and Pine Ridge Equestrian Center of Eagle River, Wisconsin, for your help with many of the photographs! Lexi Gaffron, Kimberly Kaitchuck, Colleen Fielder, and Ella Baltus for being so helpful and for making it fun, too. You girls are great! Roger James, for help with content accuracy and for generally keeping us on our toes. Teresa James, for reading the text, even on her vacation! Dr. Mike Graper, for checking over chapters 6 and 7. Lorin Johnson for proofreading, Paulette Johnson for hours of photo editing, Joshua Johnson for his fabulous ideas, Emily Johnson for help with research, and Anna Rose Johnson for researching and proofreading. We couldn't have done this without you all. Our cheerleaders: the Chief, the Artist, the Phlorida Philosopher, and Doris Johnson; we thank you! Every person should be lucky enough to have one person who thinks he or she is practically perfect and that every idea he or she has is the best one ever. For us that's been Nonee, without whose support and mentoring, this would not be in existence. Special thanks to Miracle Welsh Ponies and Cobs of Phelps, Wisconsin, for the use of their farm and facility.

And, last but not least, thanks to our pals, B.E. and R-with-two-N's, for all of the fun.

INTRODUCTION

• •

This book was written to give readers a good starting point for learning how to care for and raise horses. Maybe you already have one horse, or maybe you are looking to purchase one. Perhaps you have several horses and dream of someday running a breeding farm and raising foals. In any case, this book offers practical information for the horse enthusiast who is striving to provide the best possible care in today's horse world. The specifics of horse care change over time as new research and knowledge helps horse owners provide the highest quality care, and we have sought to give you up-to-date information to help you in your mission.

This book will help you choose the right horse and will give advice on how to feed, house, and care for it. You'll learn how to stock your stable with the proper equipment, how to understand your horse's emotions and methods of communication, and how to communicate clearly with him in return. We'll introduce you to breeds, colors, markings, and general terms, as well as conformation. We'll discuss the options that face you if you decide to raise a foal and give you insight on breeding options. And, last but not least, we'll show you how to have fun with your horse! This book is designed to present the reader with an enjoyable experience by clearly illustrating the major points with attractive photography, and to present an overview of all aspects of horse care. We hope we have succeeded and genuinely wish you the very best in your horse endeavors.

Happy Trails!

SO, YOU WANT A HORSE...

An American Paint Horse with an attractive coat pattern holds a standard conformation pose. The American Paint has a huge following across the United States.

You've just embarked on a very exciting journey! The purchase of your first horse is a wonderful, life-changing experience and one that will bring you many years of pleasure. You will need to make many decisions when you begin shopping for a horse and it helps to start with a good, firm idea of exactly what you're looking for. The best way to do that is to start with some basic information. In this chapter, we cover the preliminary essentials: breeds, colors, markings, and conformation.

POPULAR BREEDS

There is truly a breed to suit everyone, whether you're in the market for something tiny (American Miniature Horse), something large (Oldenburg), or something in between (Arabian). Whatever your plans for your horse, there's a breed that excels at it. Let's take a look at some of the more popular breeds, as well as some of the more unusual ones, and examine which one might be the perfect choice for you.

AMERICAN PAINT HORSE
Average Height: 15 to 16 hands
Origin: United States
Association Website: www.apha.com

Characteristics: The eye-catching coat colors of the American Paint Horse are the trademark of this colorful breed. Its Quarter Horse heritage makes the Paint Horse particularly suited to Western disciplines. Noted for their easy-going dispositions, they appeal to fanciers of stock-type horses who also enjoy the unique characteristics of the coat patterns.

AMERICAN QUARTER HORSE
Average Height: 14.3 to 16 hands
Origin: United States
Association Website: www.aqha.com
Characteristics: The most popular breed in America, the Quarter Horse has won the hearts of countless horse enthusiasts, largely in part to the breed's agreeable disposition and fabulous work ethic. Quarter Horses are the ultimate stock horse and noted for their "cow sense."

AMERICAN SADDLEBRED
Average Height: 15 to 16 hands
Origin: United States
Association Website: www.saddlebred.com
Characteristics: One of the most well-known gaited horses, the Saddlebred has five gaits, attractive heads, and

This lovely palomino Quarter Horse is a nice example of his breed. The Quarter Horse originally got its name for its ability to excel at racing short distances, such as the quarter mile. While Quarter Horse racing still exists, the breed today is more noted as an all-around riding animal.

high tail carriage. A popular show horse for riding as well as fine harness events. Their pleasant dispositions make them a fine choice for many equestrians.

AMERICAN STANDARDBRED

Average Height: 15.2 hands
Origin: United States
Association Website: www.ustrotting.com
Characteristics: Originally developed for harness racing, the American Standardbred is still the fastest trotting horse in the world today; although the breed has also moved on to excel in many other areas. Their powerful hindquarters and natural endurance are combined with a kind personality to produce a popular, well-liked breed.

APPALOOSA

Average Height: 14 to 16 hands
Origin: United States
Association Website: www.appaloosa.com
Characteristics: In addition to its unique variety of coat patterns, the Appaloosa is a popular choice for Western riders, as its infusion of Quarter Horse blood has accentuated the Appaloosa's natural stock-horse build. Appaloosas are naturally athletic and energetic. There are many varieties of Appaloosa coloring, which include snowflake, leopard, frost, marble, and blanket.

ARABIAN

Average Height: 14.1 to 15.1 hands
Origin: Arabia

This attractive American Saddlebred is at home in his pasture. The beautifully shaped head of the Saddlebred is one of its well-known features.

The American Standardbred became famous for its success in the sport of harness racing. Here a young Standardbred stallion holds a pose for his handler.

The Appaloosa is noted for its wide range of colors and coat patterns, as can be seen here. Two very distinct horses of the same breed graze peacefully together.

Association Website: www.arabianhorses.org

Characteristics: The Arabian is widely recognized as the most beautiful horse breed in the world. Along with their delicate, dished heads; magnificent movement; and high tail carriage; they are also noted for their stamina and versatility, excelling in a myriad of disciplines.

MORGAN

Average Height: 14.1 to 15.2 hands
Origin: United States
Association Website: www.morganhorse.com
Characteristics: Very versatile, well-suited to a variety of disciplines, and known for their stamina. Noted for their excellent action, Morgans are highly successful as driving animals.

SHETLAND PONY

Average Height: up to 42 inches (10.5 hands)
Origin: Shetland Islands, Scotland
Association Website: www.shetlandminiature.com
Characteristics: These compact, hardy ponies have long delighted children and adults alike and are popular as show ponies, pasture pets, and riding and driving animals. The classic Shetland is a hale and hearty pony, not over 42 inches, while the modern Shetland (which contains Hackney, Arabian, and Thoroughbred blood) is more

The sculpted head of the beautiful Arabian is a wonderful part of his appeal, but don't overlook what useful animals they are in general.

refined with flashier movement and may be as large as 11.2 hands.

TENNESSEE WALKING HORSE

Average Height: 15 to 17 hands
Origin: United States
Association Website: www.twhbea.com
Characteristics: A gaited horse breed, which is famous for its four-beat running walk. Tennessee Walking Horses are also noted for their excellent temperaments. They are popular show mounts and enjoyed by many trail riders.

THOROUGHBRED

Average Height: 15.2 to 17 hands
Origin: United Kingdom

Morgans are powerful animals that are full of energy and perform well in harness, as well as under saddle. The breed originated in Vermont in the early 1800s and is the official animal of that state.

Association Website: www.jockeyclub.com
Characteristics: Thoroughbreds are noted for their athleticism, excelling in all English riding disciplines, as well as being the world's premier flat racing breed. Well known for their courage, endurance, beauty, and grace, this popular breed is a perennial favorite of all horse lovers.

WELSH PONIES AND COBS (SECTIONS A, B, C, AND D)

Average Height: 11 to 15-plus hands
Origin: Wales, United Kingdom
Association Website: www.welshpony.org
Characteristics: There is truly a Welsh Pony or Cob to suit every member of the family because this beautiful, versatile breed comes in four distinct sections. Every section is well-suited to many disciplines, from short-stirrup classes with children and hunter classes to English and Western pleasure and driving. The smallest section, the Welsh Mountain Pony (Section A), stands less than 12.2 hands tall and is noted for being the most beautiful native pony breed in the world. The Welsh Pony (Section B) goes up to

14.2 hands in height, and because of its more refined appearance it is a popular choice for hunter ponies. The Welsh Pony of Cob Type (Section C) is a powerful, substantial creature with a height limit of 13.2 hands. Finally, the Welsh Cob (Section D) is a massive animal with enormous bone and movement that is over 13.2 hands with no upper height limit.

A charming Shetland Pony looks up from the end of his lead-line. A long time favorite mount for children, the Shetland Pony today remains a good choice and a fun driving animal for those who may have outgrown his small stature.

MINOR BREEDS

AMERICAN MINIATURE HORSE

Average Height: no more than 34 inches (8.5 hands)
Origin: United States and Europe
Association Website: www.amha.com
Characteristics: These smallest of equines are enjoyed by many for their diminutive size. Ideally, they are a miniature version of a regular horse with everything proportionate and symmetrical. Correct conformation is important for making the "smallest possible perfect horse."

BELGIAN

Average Height: more than 16 hands
Origin: Belgium
Association Website: www.belgiancorp.com
Characteristics: These powerful animals descend from the medieval Great Horses, and are said to be America's favorite draft horse today. These massive horses possess incredible substance and bone and have long been prized for their strength and stamina as work horses.

CONNEMARA

Average Height: 14 to 14.2 hands
(13 to 15 hands in the U.S.)
Origin: Ireland
Association Website: www.acps.org
Characteristics: One of the larger pony breeds, the Connemara is a smart, athletic, versatile, affectionate breed. They commonly participate in large competitions, routinely winning at a high level against horse breeds of larger stature.

FRIESIAN

Average Height: 15.2 to 16 hands
Origin: Netherlands
Association Website: www.fhana.com
Characteristics: This robust, impressive breed is known for its high-stepping movement at the trot, a characteristic that makes it a desirable harness horse. Friesans also excel under saddle in events such as dressage. Friesians are always black and no white markings are acceptable for purebred status with the exception of a star.

The Tennessee Walking Horse is a very suitable beginner's mount because of its quiet temperaments and easy-to-ride gaits. Many Tennessee Walkers have long, flowing manes and tails, a feature some people find quite attractive.

HAFLINGER

Average Height: 13.2 to 15 hands
Origin: Austria
Association Website: www.haflingerhorse.com
Characteristics: A strong, sturdy, intelligent mountain pony breed with a charming head, the Haflinger is a popular children's pony and can ably perform a variety of tasks. Longevity is a very common trait. Haflingers often achieve lifetimes of impressive duration, with many remaining healthy and energetic as long as 40 years.

HANOVERIAN

Average Height: 16.2 hands
Origin: Germany
Association Website: www.hanoverian.org
Characteristics: Noted as a beautiful, athletic competition horse, the Hanoverian's agility makes them fantastic ridden performance animals, well known in both show jumping and dressage. Their natural balance and grace adds to this attractive overall package.

MISSOURI FOX TROTTER

Average Height: 14 to 16 hands
Origin: United States
Association Website: www.mfthba.com
Characteristics: Enjoyed by many as a trail horse, the Missouri Fox Trotter has three distinct gaits: the flat foot walk, the fox trot, and the rocking horse canter. The fluid gaits and easy disposition of the Missouri Fox Trotter make it a perennial favorite of many trail and pleasure riders.

NORWEGIAN FJORD

Average Height: 13.2 to 14.2 hands

Origin: Norway

Association Website: www.nfhr.com

Characteristics: A hardy, versatile breed, the Norwegian

A delightful Welsh Mountain Pony and her young foal pose under some autumn colors. The Welsh Mountain Pony is frequently referred to as the world's most beautiful native pony breed.

Fjord is noted for its unique coloring and primitive markings. Regarded as an ancestor of many draft breeds, the Fjord is a powerful work horse and capable of many tasks both in harness and under saddle.

OLDENBURG

Average Height: 16.2 to 17 hands

Origin: Germany

Association Website: www.isroldenburg.org

Characteristics: Originally developed in the seventeenth century, the Oldenburg was originally used as a carriage horse but is more recently popular as a sport horse for jumping, dressage, and driving. It is noted for having higher knee action than some of the other sport horse breeds.

PASO FINO

Average Height: 13 to 15.2 hands

Origin: Latin America

Association Website: www.pfha.org

Characteristics: This gaited breed's name comes from its original Spanish description Los Caballos de Paso Fino, which translates to "the horses with the fine walk." Indeed, Paso Fino enthusiasts praise the breed's leisurely, smooth riding gait, as well as the breed's delightful personality. Paso Finos are popular show horses.

PONY OF AMERICAS

Average Height: 11.2 to 14 hands

Origin: United States

Association Website: www.poac.org

Characteristics: Originally developed in the 1950s by crossing Appaloosas with Shetland Ponies, the Pony of Americas makes a wonderful performance pony for both children and adults, combining the athleticism and coat color of the Appaloosa with the small size and hardiness of the Shetland. Other influential breeds have included Welsh, Arabian, and Quarter Horse.

ROCKY MOUNTAIN HORSE

Average Height: 14.2 to 15.2 hands

Origin: United States

Association Website: www.rmhorse.com

A beautiful, athletic Thoroughbred is about to make a quick stop from a canter while casually playing. The Thoroughbred under English tack is a common sight at all levels of horse shows.

A powerful young Welsh Cob stallion demonstrates the incredible trot and endurance the breed is known for, while maintaining a quiet and easy-to-handle temperament.

An American Miniature Horse and her young foal frolic in a tranquil atmosphere. Miniature Horses are not considered a pony breed and maintain the proportions of a full-size horse.

Here a team of powerful Belgians compete in a weight-pulling competition on a cold winter morning. Many Belgians are trained and kept specifically for the task of pulling large loads for work and competition.

Characteristics: Although the registry was not established until 1986, the Rocky Mountain Horse dates back to the late 1800s. Many Rocky Mountain Horses display a distinctive chocolate-colored coat accompanied by a flaxen mane and tail. The breed's fluid, smooth gait is popular for long distance and pleasure riding, while its steady temperament and exceptional stamina make it the preferred choice of some equestrians.

TRAKEHNER

Average Height: 16 to 17 hands
Origin: East Prussia
Association Website: www.americantrakehner.com
Characteristics: These European Warmbloods were named for the farm in Trakehnen, East Prussia, where the original Trakehner horses were bred. Native horses were crossed with Thoroughbred and Arabian horses to produce a refined, strong working horse. Trakehners are excellent dressage mounts and also excel at jumping.

CROSSBRED HORSES

Now, it's quite possible that you may decide to purchase a horse that isn't a purebred and might be a cross of two or more breeds. Some specific crosses have long been valued and admired for their quality, and the unique blending of two compatible breeds can produce individuals who inherit the best attributes of both parents.

The cross of a Thoroughbred and an Arabian produces what's known as an Anglo-Arab. This crossbreed was first produced in the eighteenth century. The breeders were looking to combine the speed of the Thoroughbred with the stamina of the Arabian to produce an athletic horse with endurance.

The Welara is the cross of a Welsh with an Arabian, and though the Welara Registry was not established until 1981, the cross of Welshes and Arabians was a major part of the breeding program at Lady Wentworth's Crabbet Stud in England that began in 1922. Welara enthusiasts strive to produce the quality and beauty of the Welsh in a larger, elegant package.

Appendix Quarter Horses are actually the result of a cross between a Quarter Horse and a Thoroughbred. The goal was to combine the ideal characteristics of each breed and produce an athletic, versatile horse.

If you can't decide on a specific breed, perhaps your solution is in one of these types of crosses. Just because there isn't a specific registry involved, it doesn't mean that you should shy away from a cross of a Connemara and an Arabian, a Welsh and a Thoroughbred, or an Appaloosa and a Quarter Horse. If the horse meets all of your other criteria, try not to be deterred by an uncommon cross.

COLORS

Horse color is certainly one of the fun factors to look at when you're shopping because there's probably a color or

An attractive Haflinger is well turned out for his pleasure driving class. In addition to driving, the Haflinger is a popular mount for young riders.

This Missouri Fox Trotter shows his kind expression. The breed is popular among pleasure riders and those who spend a great deal of time on the trails.

Many people love the unique color and markings of the Norwegian Fjord, a breed whose heritage is in farm work. Today the Fjord remains a powerful animal in harness.

two that catches your fancy more than any other. Is there anything more impressive than a gorgeous horse in your favorite color? As enjoyable as it is, it's important to remember the age-old adage, "A good horse is never a bad color." Try not to limit yourself to one specific color, as you might miss out on an absolutely fabulous grey horse while you're waiting for an absolutely fabulous bay. And what about those other colors and terms, like "black roan" or "tobiano"? What exactly are they? The common horse colors and white marking patterns are covered here to give you a better idea of what's available.

BAY

A reddish-brown body color with black points (mane, tail, legs, and tips of ears), bay is a very common color in many

A nicely turned out Trakehner and his rider prepare to enter the ring at a dressage show. The crossing of Arabian with Thoroughbred blood has produced this breed that displays the good points of both breeds.

(Right) The Morab illustrated here is a cross between a Morgan and an Arabian. This horse is performing a dressage test at competition. There is a Morab registry in the United States.

breeds. Bays can range in shade from light (golden bay) to dark (mahogany bay), along with shades in the middle range (red bay). Bays are noted in folklore as being particularly hardy. Popular with traditionalists who prefer classical colors, bay has long been a perennial choice in the hunter ring. The Cleveland Bay is the only breed which is comprised of all bays.

BLACK

Popularized in children's literature by Walter Farley's The Black Stallion, black horses have long enjoyed popularity due to their striking coloring, which is completely black

Characterized by a reddish-brown body and black points (legs, mane, tail, and tips of ears), the bay has long been a favored color with horse owners. It can be a very eye-catching color, especially when coupled with white leg markings, as shown here.

Black is a self-explanatory color. The body, legs, mane, and tail are all jet black. White markings are not allowed for some purebred breeds, such as the Friesian. White markings on black purebred Friesian horses are discouraged.

with black manes and tails, occasionally accompanied by white markings. While some black horses fade to brown during the summer months, true non-fading blacks remain jet black year round, despite time spent in the sun. All registered Friesians are black, and the occasional chestnut Friesian that occurs is unregisterable. Most black foals are actually born a lighter, mousy shade of grey and darken to black upon shedding their foal coats. Interestingly enough, foals that are born jet black often go grey if at least one parent is grey.

Looks can be deceiving! This adorable mouse-colored foal will actually be black upon maturity. Black horses are often born this unique shade of grey and later shed out to black.

Genetically, buckskin and dun are not the same color, but many people use either term to describe a cream-colored horse with black points. Illustrated here is a buckskin. She does not exhibit the dorsal stripe and zebra leg markings that would probably be present if she were dun.

BUCKSKIN/DUN

Picture a cross between a bay and a palomino and you've envisioned a buckskin. A buckskin's golden coat is accented by a black mane and tail and black legs. Genetically, buckskins are produced by the same dilute gene that produces palominos, the difference being that buckskins are dilute and bay, while palominos are dilute and chestnut.

Due to the fact that their coloring is very similar, many people use the terms "buckskin" and "dun" interchangeably, but in reality they are completely different and the coloring is caused by different genes. True duns, while retaining the golden coat color and black points, may also have a primitive dorsal stripe down their back and often have zebra stripes on their legs. Duns come in many shades including red dun, apricot dun, and clayback dun. Norwegian Fjord horses are always a shade of dun.

This jet black foal will actually be grey (the same color as her dam) when mature. Some horse breeders joke about the fact that black horses are born grey and grey horses are born black.

Chestnut is usually a shade of reddish-orange but without the black points that characterize the bay. The manes and tails can be the same color as the coat or they can be considerably lighter, as shown here. This mare has a flaxen mane and tail.

A cremello horse is a palomino with a second dilution gene that lightens the coat to nearly white. Cremello horses feature blue eyes and pink skin and have been referred to historically as blue-eyed creams.

Paint or Pinto horses have large patches of white throughout their bodies. For many years they were referred to as "piebalds" and "skewbalds," but great strides in genetic research have allowed a greater understanding of the various marking patterns and more specific terms are now in use.

CHESTNUT

Chestnuts have reddish-orange body colors but without the black points that characterize the bay. The legs of chestnut horses are the same color as their bodies. Their manes and tails sometimes match their body color, but

Grey horses lighten in color as they age until they become completely white. This eight-year-old mare still retains the darker mane, tail, and legs, as well as dapples throughout her coat. As the years pass, her coat will continue to lighten and the dapples will likely disappear.

often they are flaxen, which is a very pale cream color similar to that of the mane and tail on a palomino. Chestnut is a recessive color, which means that breeding two chestnuts will always result in a chestnut foal. Many a breeder has been surprised by the arrival of a chestnut foal from two black parents, but this is entirely possible if neither parent is homozygous for black. The Haflinger breed is made up entirely of chestnuts in various shades.

CREMELLO

Occasionally incorrectly referred to as "albino" or "white," the cremello horse is actually a palomino with a second dilute gene that lightens the coat color to cremello. Cremello horses have pink skin and blue eyes and are currently enjoying a surge of popularity for use as breeding animals, due to the fact that they will pass one copy of their dilute gene 100 percent of the time to their offspring. Some breed registries discourage the production of cremellos, and cremellos were ineligible for registration with the American Quarter Horse Association until a recent rule change. Connemaras will not register cremello

The roan pattern is a modifying gene that causes white hairs across the body of the horse (these do not extend onto the legs or head of the horse). Shown here is a black roan mare. If she did not have the roan color pattern she would be totally black.

Another example of the roan color pattern is this chestnut roan. Without the roaning she would simply be chestnut. This color is often referred to as a "red roan," but chestnut roan is technically correct.

A lack of leg markings often means that the facial white is also diminished, as shown on these two fillies. Neither exhibit any white markings other than a small star.

This colt's four full stockings are accompanied by a large blaze. Interestingly enough, this colt is a genetic full sibling to the bay filly in the previous photo, which illustrates that the inheritance of white markings can sometimes be surprising!

This type of facial marking is quite extensive. It is nearly a bald face and would be termed as such if the blaze extended over this mare's eye.

A type of overo pattern, this splashed white horse exhibits only minimal color on its face and inside its ears. This is a very striking color pattern.

stallions, but cremello mares and geldings may receive registration papers.

GREY

Grey is not a true color, it is a modifier that affects the base coat and gradually lightens it until the horse is completely white. A common misconception is that all grey horses are born black, when in actuality grey horses can be born any color, depending on the genetics of their sire and dam. Some grey horses will stay at the darker shades of grey for many years (steel grey, flea-bitten grey, dapple grey), while others become white very early. Grey is a dominant gene, which means that at least one parent must be grey. A common misconception is that grey can skip a generation, when it really doesn't matter how many of the horse's ancestors are grey: if neither parent is grey, the horse cannot be grey.

PAINT

A catchall term for horses with patterns of white markings, Paint is an outdated term that has been replaced in modern times with more specific terms such as tobiano, sabino, overo, and tovero. These more detailed terms describe the specific placement and arrangement of the white markings and are more accurate than the general Paint or Pinto description. (See the section on white markings later in this chapter for more information on the various patterns of white markings.)

PALOMINO

Ideally, palomino horses are the shade of a newly minted gold coin, but they often range in color from very pale cream to dark orange, with only some palominos falling into the ideal golden shade. Palominos have very light (nearly white) manes and tails and have long been admired for their eye-catching gold color. Roy Rogers and his palomino horse, Trigger, helped popularize palominos during the 1950s and 1960s. Genetically, palominos are chestnut horses with one copy of a dilute gene, which lightens (dilutes) their coat color from darker red (chestnut) to lighter gold (palomino).

ROAN

The roan gene affects the base color of a horse by interspersing white hairs amidst its base coat color. Roans can occur over any regular base color (bay, black, brown, chestnut, buckskin, or palomino) and the resulting colors are properly described using the base color plus the word "roan" (bay roan, black roan, chestnut roan), although traditional names include strawberry roan, red roan, and blue roan.

Like the grey gene, roan is a dominant gene and therefore a roan cannot occur unless one or both of the parents is a roan. Sometimes a heavily marked sabino with lots of flecking can be mistakenly referred to as a roan, but the true roan is characterized by a dark face and legs, and a lighter body that changes color throughout the seasons, which the sabino does not. Homozygous roan (two copies of the roan gene), which occurs 25 percent of the time when both parents are roan, is believed to be lethal in most cases.

This is a beautifully marked sabino horse. The roaned and speckled edges are characteristic of the sabino color pattern, as opposed to the more crisply defined edges of the frame overo or splashed white.

Unlike Overo Lethal White Syndrome, which is fatal after the birth of the foal, lethal roan occurs early on in pregnancy, typically before the pregnancy is even confirmed. Very few, if any, homozygous roans are known to exist.

WHITE MARKINGS

While there are a few purebred associations (such as Exmoor Ponies) that discourage or prohibit horses with white markings of any sort from registration, most associations allow horses with white markings, which can range in degree from a simple star and sock, to the unusual and eye-catching patterns of Appaloosas and Paint Horses.

An example of the tobiano color pattern is shown here with the more conservatively marked face and four white legs. These are typical characteristics of the tobiano.

Each breed registry has its own guidelines regarding how much or little white is allowed on a purebred animal.

Research has shown that the amount of facial white is somewhat correlated to the amount of white on the legs. Generally speaking, a horse with a generous white blaze will have high leg white, and a horse with a small white star is more likely to have legs with little to no white markings. Additional research has shown that the horse's coat color can also have an effect on the extent of its white markings, with black horses typically exhibiting less white markings than bay horses, and chestnuts having more white markings than either bay or black.

Facial markings can range from a small star to a blaze or bald face, and leg markings can range from a white coronet all the way up to a full white stocking or onto the knee or hock.

In recent years, there have been many developments in the understanding of white markings, their inheritance, and the specifics of particular patterns.

OVERO

The term "overo" is generally used to describe a Paint horse that is not tobiano. They generally have a lot of white on their heads, and often have blue eyes. Overo coloring includes three different patterns: frame overo, sabino, and splashed white overo:

FRAME OVERO

The frame overo is named for its unique pattern of white markings, which appear to be framed by the darker coloring of the base coat color. White on the face is very common with frame overos, as are blue eyes. The legs can be dark. Frame overos are the color pattern most often responsible for Overo Lethal White Syndrome (OLWS), which is a fatal condition when a foal is born white and cannot survive due to a colon that does not function properly.

A great deal of research has been conducted on the genetics that cause OLWS and studies have found that foals with OLWS have two copies of the lethal gene, having inherited one copy from each carrier parent (each of whom also carry a normal gene, and thus are personally unaffected by the lethal gene). Research has shown that overo horses can carry either two normal genes or one normal and one lethal and remain unaffected. It is only fatal when two copies of the lethal gene are present. However, care should be taken when breeding an overo horse who is a carrier of a lethal gene to avoid the possibility of producing an OLWS foal. Genetic testing can be done to determine which genes an overo horse carries.

SABINO

Characterized by white markings that are speckled or roaned, sabino is found in many breeds, including Welsh Ponies, Thoroughbreds, Clydesdales, and Quarter Horses. Sabino is the Spanish word for "pale or speckled." In cases where the sabino pattern is minimally expressed, the horse may have only stockings and a blaze, perhaps accompanied by some speckling or roaning along the flanks. These minimally marked horses have the potential to produce foals with more generous markings, especially when crossed with another sabino. Sabinos that are more extensively marked may have roaned patches of white on their bodies, under their chests, or on the jaws. Occasionally, a minimally marked sabino may display little outward characteristics of the pattern (perhaps a small star, and dark legs) but will have the giveaway white flecking throughout his coat.

SPLASHED WHITE OVERO

Another very unique pattern, the splashed white is quite different in appearance from the sabino. Splashed whites have markings with crisp edges, not roaned as the sabino. It has been noted the splashed white looks as if it has been

When studying the conformation of a horse, examine him closely from both sides to get an overall view of his balance and length of rein and tailset. Look from the back and front of the horse as well to observe the straightness of his legs. Be sure to examine him from all angles to ensure the best possible understanding of his strong and weak points.

The conformation of foals can be quite difficult to judge, as they are still growing and changing and going through stages that can seem to make things unbalanced or incorrect. Usually by the time a horse is three years old, his conformational development is complete.

dipped in paint, with the white markings being more heavily concentrated on the lower portions of the body and the head.

TOVERO

Tovero is a term recognized by the American Paint Horse Association and is used to describe horses that are a combination of more than one color pattern. Essentially, tovero means that the horse displays the characteristics of a tobiano and one of the overo patterns. Tovero is probably one of the more difficult patterns to pinpoint, as it can be exhibited in so many different forms.

TOBIANO

Typically characterized by white legs and spots, the tobiano is noted for having a head with minimal white and dark

eyes, as opposed to the overo varieties. Tobianos often have ink spots of dark intermixed in their white patterns.

CONFORMATION

The importance of correct conformation is sometimes overlooked by the novice horse owner. After all, if the horse has a good disposition, is well-trained, and is healthy and sound does it really matter if he's pig-eyed or has crooked legs? You can enjoy years of pleasure with a long-backed or straight-shouldered horse, and a calm, steady trail horse is good at his job even if his ears are too big. So why does conformation matter at all?

Correct conformation, in addition to being aesthetically pleasing to the eye, is fundamental to optimum performance. Good conformation makes a good performance horse perform even better. The basics of correct conformation are built upon the necessary attributes of the ideal equine.

If you're planning to show your horse, good conformation is of vital importance. Halter or breeding classes are judged on how well each individual horse conforms to the ideal picture of the perfect horse. Obviously, each breed standard has subtle differences that place emphasis on particular parts of the overall horse, but most points of good conformation are generally universal throughout all breeds. But what if you're only planning to show in performance classes? Does conformation still matter? The answer is another yes. Proper conformation allows your horse to perform at his maximum capability with less chance of injury.

If breeding is in your plans, you will also find conformation to be of the utmost importance when choosing breeding stock. Perpetuating faulty conformation will not only result in foals that are of lower quality and of lower value, but also makes them less in demand when you are ready to sell them. Only individuals of good quality, breed type, and conformation should be used as breeding animals. Those with major conformation faults should not be chosen for a breeding program.

So, what is good conformation, and how does one learn to recognize it? There are many wonderful books on conformation and how one can learn to recognize it, as well as videos that can also be very helpful in visualizing the differences between correct and incorrect conformation. Judging clinics are also great sources for learning about conformation, even if you aren't interested in becoming a judge. Auditing a judge's clinic is a great way to increase your knowledge, and you can typically audit a clinic for a nominal fee, particularly when considering the excellent amount of information you can gain.

While it would be impossible to thoroughly cover all of the aspects of good conformation in this chapter, some of the major points are highlighted.

A properly conformed horse is balanced. Each portion of the horse should be a balanced match with the other portions so that no single part seems unduly out of proportion. The

Parts of a Horse

Poll, ears, crest, withers, back, loin, croup, dock, flank, thigh, stifle, gaskin, hock, ergot, hoof, heel, coronet, pastern fetlock, cannon, knee, forearm, chest, shoulder, throat latch, jaw, muzzle.

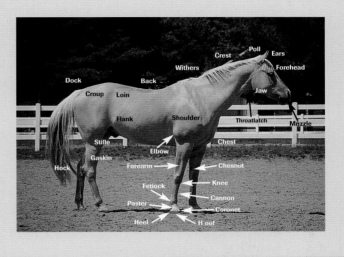

front should be equivalent to the mid-section and hindquarters. The amount of bone should be adequate for the size of the body and the ratio of body and leg should be even.

The head should be attractive with a pleasant expression. The eyes should be bold, not sunken, and prominent on the face. The ears should be well placed and not too far forward or back. The length of the ears is dependent upon the breed and size of the horse but should not be so large that they seem out of proportion in regard to the rest of the horse's head. There should be adequate width between the horse's eyes, and the nostrils should be large and open (not narrow). Parrot mouth (misalignment of the teeth) is a serious fault and should be avoided. The throatlatch should be well defined and not thick.

The neck should be of an appropriate length in order for the overall appearance of the horse to be balanced. Ideally the top side of the neck should be one-and-a-half to

Common Conformation Terms and Definitions

Bench Knees   Offset knees; the cannon bones do not come out from under the knees

Bowed Hocks   The opposite of cow hocks; these are hocks that are bent away from each other

Bull Neck   A short, thick neck

Calf Kneed   When viewed from the side, knees that are set too far back

Cow Hocks   Hocks that are bent toward each other rather than straight

Ewe Neck   A neck that is set upside down; sunken on the top and protruding on the bottom

Goose Rump   A steep croup with a low tail set

Mutton Withers   Flat withers with fat pads on either side of the withers, which make saddle fitting very difficult

Parrot Mouth   Overbite; misalignment of the teeth

Pig Eyes   Eyes that are too small and not prominent on the face

Pigeon Toes   Also known as toeing in; the hooves are turned toward each other and are not straight

Post Leg   A hind leg that is too straight

Roach Back   A deviation from the ideal shape of the back; convex

Roman Nose   Convex profile of the head; the opposite of a dished head

Sickle Hocks   An overbent hind leg; hind feet set forward under the body

Splay Footed   Also known as toeing out; the hooves point away from each other rather than straight

Swan Neck   A neck that is too long

Sway Back   More common in aged horses, it is an excessively low back much lower than the points of wither and croup

Wasp Waisted   Narrow waisted; decreasing abruptly in appearance from the rest of the midsection

A quality broodmare with a kind eye and pleasant expression, in addition to her excellent conformation, will go a long way toward the beginning of a successful breeding operation.

two times longer than the underside of the neck. The neck should flow smoothly into the withers and into the back without a depression in front of the withers. The slope of the shoulder is an important characteristic as it dictates the length of the horse's stride. A straight-shouldered horse generally has a shorter, choppier stride than a horse with a properly sloping shoulder. A horse with a sloped shoulder has better freedom of movement, and this results in a longer, smoother stride. The back should be short because a short back is fundamentally stronger than a long back. Long backs are a common fault, but many people consider this fault to be less serious in broodmares, as it gives them more room to carry a foal. The hindquarters should be deep with the croup not being too steep.

The hind legs should be straight when viewed from behind with the hocks turning neither inward nor outward. The pasterns should be of moderate slope, approximately 45 degrees, and deviations on this angle are viewed as incorrect. Similarly, the front legs should also be straight and the hooves should not turn in or out. The horse's hooves should be round, not narrow.

The importance of good conformation shouldn't be underestimated. While you're horse shopping, if you're at all unsure about a particular component of a horse's conformation, you can ask the veterinarian to evaluate it during the pre-purchase exam for an educated opinion about any conformational faults and the possible effect on your future plans for the horse. Seek the advice of a friend who is knowledgeable about horses to help you determine whether the horse's conformation is correct, and if not, which points could use improvement.

It's good to remember that as important as conformation is, there is no perfect horse. Each and every horse on the planet has a body part that could be improved upon, so don't despair if you can't find a horse without a slight flaw of some kind. As you talk to other people about conformation, you will find that most people have particular points of conformation that they are especially passionate about. Some people can't stand a horse with cow hocks, and others are very particular about having nice, round hooves. Other people might be very particular about the length of the back. Obviously, all points of conformation are important, but there may be some that are more important to you than others.

CHOOSING YOUR HORSE

When you think you've found the perfect horse, it may not matter to you if he is registered or not. Try to avoid falling in love too quickly with a horse and making a rush decision.

Now that the different breeds (popular and unusual alike) have been discussed, you probably have formed some idea of which breed appeals to you and which ones are simply not right for your needs. You've studied the essentials of correct conformation and can spot a cow-hocked or ewe-necked horse the moment you see one.

Equipped with this knowledge, you've decided that you're ready to proceed on to the next step—buying your horse. Wait just a moment! There are still several factors to consider and decisions that must be made before you make your final choice.

WHAT'S RIGHT FOR YOU?
REGISTERED VS. GRADE

If you've already settled on a specific breed to purchase, the choice between a registered horse and a grade (unregistered) horse may be irrelevant. You may have already decided that a registered Quarter Horse is the only option you're interested in so that you may show at Quarter Horse shows and possibly raise a purebred, registered foal in the future. In that instance, your choice is already made. But what if you're only planning to show locally at open shows and you have no plans to breed? Do you really need a registered animal, or would a grade horse be perfectly suitable?

Backyard family horses that will be used for pleasure and enjoyment can come in all shapes and sizes and don't need to be accompanied by registration papers. A grade horse can often be less expensive than a registered one, which is appealing to many first-time horse buyers. Resale values on registered horses tend to be higher than those of grade horses, which is an important consideration if you think you're eventually going to move up to a different horse.

For many disciplines, a registered horse is not necessary in order to compete, but breed shows typically require registered animals. If you have any inclination to raise foals in the future, using registered stock will probably result in foals that have a higher dollar value than foals from grade

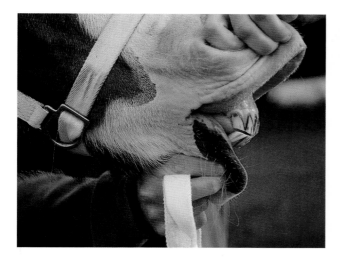

Check the teeth on any horse you're considering, especially if she is a grade horse. An experienced horse person can give you a general idea of a horse's age based on her teeth. This long-toothed mare is 24 years old.

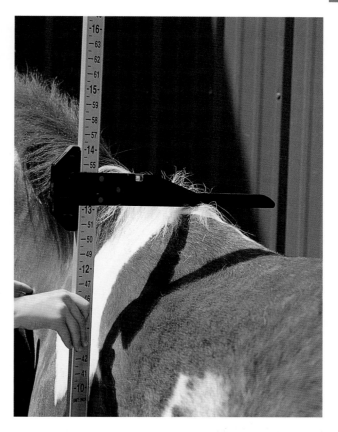

It might be wise to get up close to animals of various heights to get a feel for what seems suitable for you. The length of gaits and general feel of riding change depending on the size of the animal.

parents. There are people who feel that only registered animals should be used as breeding stock, but many others feel that they are able to achieve satisfactory results with grade stock, as in the case of those breeding for the sport pony or sport horse markets, which are often produced using grade or crossbred animals. The expenses to keep a horse are the same whether it is registered or not, so many people feel that it is worth the additional purchase cost to increase the value of their investment. If you're looking to purchase a registered horse, it's always important to ask if the seller is in possession of the horse's registration papers. Even if a horse is advertised as registered, it's possible that the registration papers are lost or were never transferred from a previous owner. In these instances, registered animals with no proof of registration are of no greater value than a grade horse if the registration cannot be provided.

SIZE

Size is an important consideration when choosing your horse. Some people may discourage you from purchasing a horse that is too small for its rider, yet many people tend to overlook the opposite and end up purchasing a horse that is too big for them, which is every bit as unsuitable. One common mistake is to assume that you need a larger horse than you really do. While a 14.3-hand horse may be considered small, it's still a very large animal and big enough to be intimidating to a novice horseman. There is virtually no practical difference in the athletic capabilities between horses that are within a few inches of each other in height.

It can be difficult to visualize the true size of a horse in terms of height and weight. If you are at all unsure, try visiting a boarding/riding stable and ask to be shown horses of various sizes so you can get close to each one and determine what height you are most comfortable around. Many people (particularly children) will pick an arbitrary size for their potential horse ("Must be at least 16 hands" or "Absolutely no smaller than 15 hands") without having any practical idea of what that size represents or what they really need.

How to Measure a Horse

The height of a horse is measured in hands. One hand equals four inches. The term "hand" dates back to the time when a man estimated the size of a horse based on the width of his own hand, which he could use as a gauge when examining a horse.

When measuring a horse, bring him to a safe, level, firm surface, such as the barn aisle. Don't measure him in his stall as the depth of his bedding (which he is standing on) may throw off the reading. Have him stand fairly square, especially in the front. It helps to have someone else hold the horse while you measure. A specialized stick, available in most tack shops, will give the best reading. Measure from the ground to the point at the top of the withers, while taking care to keep the stick perpendicular to the ground. Some measuring sticks have a small bubble level that will assist you in taking an accurate measurement. The top of a horse's withers can be difficult to judge at times, so it may help to have someone encourage the horse to lower his head and neck as this will make the withers stand out more prominently. Take a few different readings if you're having trouble and average them out.

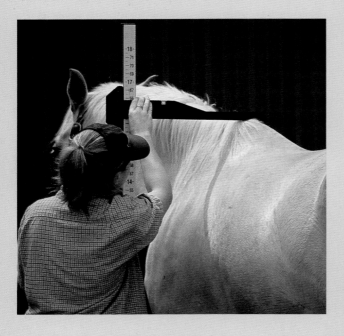

If you're purchasing a family horse to use for pleasure riding on trails it really makes no difference whether you purchase a 14-hand pony or a 15.3-hand horse, except that the pony would have the advantage of being a more universal size for a family of various experience levels and abilities. Quiet, smaller ponies are always in high demand, as there is no better match for a beginner child than a sensible, well-trained pony of 12 to 13 hands. Young riders can learn the basics of handling, saddling, bridling, brushing, picking out hooves, mounting, and riding on an animal that is precisely sized to fit them. Unfortunately ponies have an undeserved reputation for being stubborn and lazy, but this stereotype is largely exaggerated. Well-trained ponies with good dispositions are a marvelous choice for a family.

MARE, STALLION, OR GELDING?

Your decision to choose a mare, stallion, or a gelding will depend upon your plans for the horse. Obviously, if you're starting up a breeding operation, then a gelding is not going to work for you. However, if you're looking for a good, family horse that will take the kids on trail rides, then a gelding may be the perfect choice.

Geldings are the preferred choice of some horse owners. Geldings are castrated males and are noted for their steady temperament, because they are not as prone to hormonal mood changes as mares and stallions can be. Geldings can sometimes be less expensive than mares or stallions, especially if you purchase one that is young or has had limited training.

Experts generally agree that a stallion is not a wise choice for the novice horse owner. Stallions can be unpredictable, strong, and require a confident, capable, experienced

A gelding will often provide a child with a long-term, steady mount who will probably be relaxed and consistent in various situations and can be a great choice for your family.

handler. While there are certainly stallions of all breeds that have gentle, quiet dispositions and can be handled by most members of the family, they are the exception rather than the rule.

Mares are also a good choice for the average owner. The majority of mares are as easy as geldings to work with and be around. Some people do complain of mare-ish behavior when the mare is in heat. Acting mare-ish can include an inability to concentrate and a change in disposition (suddenly placid and sweet or suddenly cranky), which lasts for the duration of the mare's cycle (5–9 days). It must be noted, however, that this is not an across-the-board phenomenon and many mares display no outward signs or changes when they are in heat. Again, if you're thinking of doing small-scale horse breeding, mares are a necessary component. Some families will choose a quality mare for their children to ride and enjoy and will later use her as a broodmare. Thus, it's important to keep future possibilities in mind when choosing a horse. If you think you might possibly use a mare as a broodmare in the future, make sure that she is of good quality, conformation, and type for her breed, in addition to the other attributes that you're considering for her present use (good disposition, trainability, and soundness).

Many horse owners choose to own a mare and enjoy her company for years, which speaks highly of her abilities and temperament. The choice between purchasing a gelding or mare does not have to be difficult because either option is a good one.

While it may be tempting to purchase an adorable foal for your child, the idea of letting them grow up together is usually misguided. For a child's mount, experience and temperament are more important characteristics to consider.

It's possible that a mare will retain her value better than a gelding. In the event that a gelding becomes injured (to the point that he is no longer sound enough to ride), his value greatly decreases. However, a mare in the same situation can go on to a successful second career as a broodmare, producing foals and retaining her value for a longer period of time.

Some owners feel that mares have more personality than geldings do, try harder while working, and are more sensitive to their riders. On the other hand, some people feel the exact opposite, that geldings are more consistent and put in a better day's work than mares.

Finally, it's important to remember that some people have a personal preference. Some people love mares and would never have anything else. Others are smitten with geldings and love their steady personalities. Some experienced horse owners prefer owning their own stallion and enjoy the challenges of maintaining and promoting him, though again, this is not recommended for a novice or beginning horse owner.

Understanding these differences is vital to making the right choice for your situation and having a good relationship with the horse you purchase. A good rule of thumb is to remember that if you're a total novice with horses, a gelding is a good choice. It's very rare to come across someone who believes they are a total novice with horses, but we know you're out there. When it comes to horses, it's always good not to overestimate your ability.

AGE

Foals are undeniably adorable and many people feel irresistibly drawn to their charm. While it's natural to want something so adorable for your very own, it's good to think through your plans and goals for the horse before you purchase one. If you're looking to ride, whether it's pleasure riding or showing, you must keep in mind that it will be several years before your foal is old enough to be started under saddle, and then even more time must elapse before he will be trained enough for a beginner. You also must consider whether you have the expertise to train a young horse. If not, you'll need to factor in the cost of training. A better option might be to purchase a well-trained mare that you can use in performance right away, which will give you the option of breeding her later.

Young horses appeal to some buyers because they can be much less expensive than a mature, trained horse, but it's vital to consider the amount of time and money that must be invested in a young horse. There is the cost of raising the horse until he is of an age to begin training, the cost of the

An older horse who has spent many miles under saddle and has years of patient experience with all types of riders is an excellent choice for a young rider. This type of horse can still give a child years of safe, pleasant moments.

This is an easy horse to handle and have around the stable. She calmly and obediently follows her handler back to the barn after being easily caught in her pasture.

training itself, not to mention the time and attention necessary to make sure that your young horse is well-handled and raised correctly.

Experienced horse owners often enjoy the task of selecting a young horse (yearling, two-year-old, or three-year-old), raising it, training it, and reaping rewards later on. But this is a complicated process that requires expertise and patience and is not recommended for the person choosing his or her first horse.

Most experts agree that the first horse for a novice horse owner should be at least eight years old. While there are certainly horses younger than eight who might be a perfectly suitable match for a novice, it is typically advisable to purchase a horse that is at least five years of age, and preferably over eight.

Don't overlook an older horse, especially if you're a novice. A horse in its late teens or early twenties can still give you several years of enjoyment and you'll benefit from purchasing a horse with many years of prior experience. However, an older horse may be more prone to health issues so make sure you get a pre-purchase physical exam before you decide to buy the horse.

Keep in mind that age is not a foolproof guide to finding a quiet, well-trained first horse. A well-trained, quiet five-year-old would probably be a better choice than a 12-year-old who was broke to ride at three years old and then turned out to pasture for the last nine years. While age can be a good gauge, it is not the only aspect to consider. It's wise to look at the whole picture before making your decision.

TEMPERAMENT

Considered by many to be the single most important characteristic when choosing a horse, temperament is a vital consideration when you're horse shopping. For a novice

An experienced horseman or trainer will be of invaluable assistance to you if you are searching for a horse to perform a specific job. Find the help of someone who understands your desired discipline to help find a horse whose attributes are geared toward this type of work.

A more difficult situation is a young filly who resists the pressure on her halter and attempts to pull backward. It may take some careful handling before she learns to lead properly.

owner, a quiet, sensible, well-trained horse is worth its weight in gold. You will want to shop carefully to make sure that you don't buy a horse with any temperament issues.

Horses have personalities. I personally prefer a horse that "thinks" and has an endearing personality (friendly, happy to see you), while other people prefer horses that mind their own business. Others like a sassy horse.

At the risk of over-simplifying, you can generally classify horses into two categories: ones that are easy to live with and the troublemakers.

CHARACTERISTICS OF THE EASY GOING HORSES:

• Can be safely turned out to pasture with other horses.
• Can be bathed, clipped, tied, loaded, and dewormed without breaking your bones or theirs!
• Don't chew wood and won't chew your barn down.
• Are happy eaters.

• Are easy to catch.
• Are "bombproof," unflappable, not easily spooked by everyday objects, such as a water bucket.
• Would never dream of biting or kicking.

TROUBLEMAKERS HAVE CHARACTERISTICS THAT CAN INCLUDE:

• Fight with other horses at pasture; a bully-type personality.
• Require twitching or sedation to bathe, clip, and deworm; may also be difficult to tie or load.
• Chew wood/cribbing.
• May be nervous eaters, suspicious of any change in diet.
• Are hard to catch; you haven't touched him since the day you bought him, or the only view you ever get of him is his backside.
• Are easily spooked; the kind that jumps if you move your wheelbarrow to a new spot.
• May bite or kick.

These are obviously generalizations and you may or may not be bothered by these distinctions. Additionally, the characteristics above only describe basic temperament. You

Purchasing a horse can include a sizable investment of time and money. There will be time spent talking on the phone with sellers, time and money spent traveling to look at horses for sale, time spent searching advertisements, and time spent researching information. The time and money spent are a wise investment and will allow you to reap the rewards when you find the perfect horse.

will want a patient, gentle, kind, quiet horse for your family, and without these fundamental characteristics, your search should continue until you find a horse with these qualities.

SUITABILITY FOR THE JOB

If you have a specific discipline in mind that you'd like to pursue, such as dressage, driving, or barrel racing, you'll want to find a horse that has natural ability, the proper build, and training in that area. It would be a poor decision to purchase a horse trained and proven in the hunter ring if your goal is to do barrel racing. Similarly, if you're looking for a broodmare for a breeding program, you wouldn't want to purchase a 20-year-old maiden mare (one who has never had a foal).

Specific disciplines are often associated with certain breeds. For instance, Quarter Horses and Paints are well known in Western riding, while Thoroughbreds and Warmbloods often excel in English disciplines.

Overwhelmed by the myriad of available horses? Before you start looking, begin with a detailed idea of what you're looking for. Only respond to ads that closely match your specifications.

Although you might be lucky enough to find the horse of your dreams by driving around town and seeing a sign, chances are that you will have to do a little searching before you find exactly what you're looking for.

that person if he or she will give you specific things to look for in a horse that will be suitable for its job.

WHERE SHOULD YOU LOOK?
BULLETIN BOARDS

Stop by any feed store, tack shop, or show barn and you won't be able to resist taking a peek at the bulletin board. Ads, photos, flyers, and phone numbers are available for the horse shopper to peruse. Advertisements are typically limited to a short period of time so you can be sure that the ads you're viewing are relatively current and local. If you can't find exactly what you're looking for on the board, you can post a "wanted" sign detailing your horse specifications, price range, and contact info. Bulletin boards tend to get a lot of exposure, so someone may see your sign and contact you with a possibility. Keep checking the bulletin board as new ads are posted often. Even if some of the ads aren't applicable to your current search, you can still enjoy looking over the stallions at stud, puppies for sale, used saddles, outgrown show clothes, and other horse items while you search for the horse of your dreams.

If you're purchasing a horse for a horse-crazy child, he or she may not care what the horse is or does, just as long as it has four legs, a mane, and a tail. However, as a parent, you must be the voice of reason. Explain to your child that he or she will be much happier with a horse that is suitable for its intended discipline. Find an expert (trainer, breeder, or judge) in the discipline that you're considering and ask

When purchasing a broodmare, you will want to evaluate the quality of her offspring, in addition to the mare's own conformation, breed type, and movement. Ask the owner for photos or video of her previous foals so that you can determine whether she is producing the type and quality that you want. As you can see, this mare has reproduced herself very closely in her foal.

WORD OF MOUTH

Spreading the word to your horse friends that you're shopping for a horse can be enough to get you some good leads. Trainers, veterinarians, farriers, and feed-store owners can all be good sources of knowledge about what is available locally and might be able to point you in the direction of a suitable choice. In some instances, your friend may know the horse personally and be able to give you first-hand information on whether or not it would be appropriate for you. Trainers can be excellent sources of contacts in your area and can give insight on a particular animal based upon their educated opinion. It has been said that the best horses are never actually listed for sale because they are sold quickly through word of mouth before they have been on the market long enough to appear in ads. Sometimes sellers will be more inclined to negotiate the asking price if they sell the horse quickly through word of mouth before they have had to spend any money on advertising.

INTERNET

The world is at your fingertips when you're horse shopping on the internet. Horse classified websites reach thousands of shoppers every day and offer a vast selection of horses and ponies for sale. You can search for specific breeds, certain sizes, or particular colors, or you can search by bloodlines, training level, attributes, location, or price. If the ad doesn't have a photo, you can contact the seller to request one, which he or she is typically able to send via email. In addition, organizations and associations usually have locations on their websites for advertisements from their members that you can browse.

If you are searching for a particular breed or only wishing to look locally, a search on Google can bring up dozens of possibilities. For instance, if you type "Welsh Pony, mare, Wisconsin, for sale" into a search engine, it will bring up several links to classified sites with Welsh Pony mares for sale in Wisconsin, as well as breeders of Welsh Ponies in Wisconsin and links to Welsh stallions at stud. There is a lot of information for the shopper! Similarly, a search for "Appaloosa gelding, for sale, Texas" will bring up numerous classified listings that meet the criteria, including breeders

Certain breeds, such as Warmbloods or sport horse and pony registries, hold inspections to evaluate athletic ability, breed type, movement, and conformation. Horses that are approved by these registries are often branded so ask for inspection scores if you're purchasing a horse with a registry brand. If you have to transport a branded horse across state lines, you may need to have him brand inspected prior to travel.

of Appaloosas in Texas and message boards where Appaloosa enthusiasts meet to chat about their breed and list horses for sale. If you want results that are even more specific, try adding in other attributes, such as registered, bombproof, or sound.

Since you might not find exactly what you're looking for on the first try, you can bookmark a particular search on an internet classifieds site and check back every few days to look at the new listings. It saves time and allows you to only look through the ads that interest you.

Breeder's websites can be particularly helpful because they often contain a wealth of information for the buyer to read. In addition to the information and photos on the

It can be difficult to judge a horse with 100 percent accuracy when shopping by video, but purchasing him from this method can allow you to find a very specific animal that may not be readily available from your location. Be polite and return the tape or disc to the seller when you're through looking.

horses for sale, you can also look at their stallion and mare pages, view show records, explore the farm's history, and see photos of the horse's relatives. All of this information can give you a good feel for the areas that the breeder's stock excel in and help you determine whether they are producing the type of individual you're seeking.

MAGAZINE CLASSIFIEDS

Flip to the back of any horse magazine and you'll find a selection of classified ads. Most are arranged by breed so if you have a specific one in mind you'll be able to quickly locate the ads that are pertinent to your search. Major newsstand magazines typically have classifieds that are geared toward offerings from breeders, as opposed to horses for sale from single horse owners or discipline-based ads. Magazines representing specific disciplines (dressage, driving, hunters, Western) will have classifieds that are strongly based upon their readership's interests. Regional or state publications are terrific for helping you locate breeders and sellers in your area. Most states have a horse publication of some sort, as do most breed associations, so if you are looking for a specific breed or particular location, these publications can be a vital asset in your search.

One important thing to keep in mind, whether you're looking at magazine or internet ads, is the hidden meaning

Here a mare and foal graze peacefully on a farm that specializes in a specific breed. If you know which breed you are interested in pursuing, a farm such as this might be a great place to look.

of the phrases in the ads. Obviously, the ad is going to highlight the horse's positive points and will usually not mention anything negative. However, I've personally learned to be leery of any animal labeled "smart," or "sensitive," or anything that "needs an experienced rider/handler." These can be polite ways of saying that the horse is jumpy, nervous, easily upset, pushy, or possibly dangerous and not a good choice for the novice owner.

BUYING FROM VIDEO

If you're searching on the internet, it's very possible that you may find the perfect horse halfway across the country. It's also possible that you will find one only a few hours away. But either way, you will want to have a little more information before committing to making the trip to see the horse in person. In this case, you can request, or the seller may offer, a video.

Videos of horses for sale can vary. Some will be short and only a couple of minutes long, while others will contain a half-hour of footage of the horse in various situations. Obviously longer is better as it will give you the most thorough evaluation of the horse you're considering. Much of the video's contents will depend on your intended use for the horse. If you're purchasing an older mare as a prospective broodmare, you will hopefully receive footage of her standing, untacked, to evaluate her conformation. You would also want to see footage of her being led to and from the camera at a walk and trot with a clear view of her legs to verify that they are correct structurally. This would be followed by footage at liberty to see her movement. If the mare has already produced foals it would be ideal to receive footage of them to get an idea of whether she reproduces herself and if her foals are higher or lesser quality than their dam.

Purchasing a trained horse is more complicated because you will want to see footage of the horse being saddled; lunged; ridden at the walk, trot, and canter; and footage showing a willingness to halt and back. If you're looking for a trail horse, it would be helpful to see the horse out on the trails. Watch for how he reacts to strange situations, spooky objects, and unexpected sights. Similarly, if

Horse Classifieds on the Web

www.acmehorses.com
www.agdirect.com
www.dreamhorse.com
www.equine.com
www.freehorseads.com
www.horsetopia.com
www.ponyworld.net

you're looking for a prospective hunter, you'll want to see how the horse navigates a course of jumps, his form over fences, and his style and regularity of gaits.

Some sellers will have the video already prepared in advance, in which case you won't have a choice as to what footage is offered. If the seller will make the video exactly to your specifications, you can request that certain segments be included. Ask to see footage of the horse being handled and led, standing tied, and other footage at liberty.

Watch the video several times over the course of a few days to give you time to really examine the footage you have been provided. If possible, have an experienced friend evaluate the horse. He or she may notice something that you did not. Don't purchase on impulse. Even though the horse may seem absolutely perfect, you should never let excitement prompt you into making a rush decision. Take your time, think it over, ask more questions if necessary, and then make your decision. What may seem like the perfect choice on Tuesday morning could turn into an "I don't think he's quite right" decision on Thursday after you've thoroughly weighed your pros and cons. Try to avoid being swayed by seller tactics because he or she may try to pressure you into making a quick decision by informing you that there are other buyers ready to buy the horse. Some sellers use these tactics to push you into making a decision before you're ready. Don't rush! Making the right

When visiting a potential purchase in person, it's a great idea to bring along an extra pair of eyes and ears. Make sure it is someone you can discuss the horse with as you observe him perform in his normal environment.

decision is more important than making a quick one. You don't want to regret your choice.

Some people would never consider purchasing entirely from a video without having seen the horse in person, while others have had excellent success with their video purchases. It really is a personal decision, and will depend on what you are comfortable with. It's your money and your horse so you must feel satisfied with your decision. If you do your homework, ask a lot of questions, and thoroughly research your purchase, you should have no reason to be disappointed when your new horse steps off the trailer and you meet him for the first time.

LOCATING AND BUYING FROM A BREEDER

If you're in the market for a Quarter Horse, Thoroughbred, or Arabian, you should have no trouble locating a reputable breeder within a reasonable distance of your location. However, if you're in the market for something a little less common you may have to do a bit of detective work to find the breeder you're looking for.

The best place to start is with the breed's association. Many association websites have breeder listings compiled by state so you can quickly locate those closest to you. Otherwise, a quick phone call to the association office can put you in touch with breeders in your area who may have

something to offer. An internet search can also uncover information on local breeders. Simply type in the breed of your choice, followed by the state where you live.

In the case of a rare or unusual breed, it's very possible that you won't be able to locate a breeder in your backyard. Obviously, it's easier if you can go directly to the breeder's farm and look over his or her stock in person, but if the breeder is across the country, you will have to either shop by video or arrange a visit. While this may seem more complicated than necessary, it is a viable option for people who are interested in particular breeds, specific bloodlines, or something they are unable to find in their area.

By buying from a breeder, you will have access to information about the horse's sire, dam, siblings, previous history, as well as specific information about the individual you are purchasing, such as "horses from this line tend to mature quickly" or "this mare's foals are the sweetest." This type of valuable information is typically unavailable when you are purchasing from a backyard owner, from a sale barn, or through an auction.

MAKING YOUR PURCHASE
GOING TO LOOK IN PERSON

If at all possible, you will want to see the horse in person before you make a decision to purchase. It's a courtesy to the seller to be certain that you are seriously interested before making an appointment to see the horse. If you really think that you want a registered Quarter Horse mare and the horse you saw an ad for is an Arabian gelding, think carefully before you bother the seller with setting up an appointment. However, if you're reasonably sure that you are interested in the horse, proceed with the appointment. It goes without saying that you will want to be on time. You're taking time out of someone's busy day and you'll be starting your appointment on the wrong foot if you arrive late.

It's a good idea to bring someone else with you. Two heads are better than one and you'll each notice things that the other one might miss. If possible, bring an experienced horse person or trainer who is familiar with your level of experience and goals for your horse. One of you can take pictures or videotape the horse for later reference. While it's perfectly acceptable to bring along a friend, it's not courteous to bring a car full of people or dogs with you.

You will want to examine the horse thoroughly yourself to determine its general state of health. Check legs and feet and look for bumps, lumps, and blemishes. Watch the horse moving from the front, back, and side; check for possible lameness; and evaluate the horse's overall conformation, particularly his leg conformation. (See Chapter 1 for information on correct conformation.) The horse should have an alert expression, appear interested and bright-eyed, be at an appropriate weight, and have a healthy, shiny coat. If the horse is not registered, try to verify the horse's actual age by checking his teeth. Observe the horse's temperament during handling, while being saddled, and while being tied. Take note of how he behaves in his stall, whether he paces or seems agitated or if he is relaxed and quiet. Some horses are nervous when they're

Questions to Ask Before Buying a Horse

- Describe his disposition. Is he good with other horses?
- Does he have any bad habits or vices?
- Does he have any health issues, major illnesses, or injuries? Has he ever been lame? Prone to colic?
- Does he react to being loaded, clipped, bathed, or tied?
- Has he been shown? Any championships or titles? If registered, has he won points at breed shows?
- Has he had any professional training? How much?
- How is he for the vet and farrier? (If this answer is "I don't know," this may be a red flag, as you may begin to question the care that the horse has been receiving.)
- How is he used to being kept (in a stall, at pasture, or both)?
- How long have you owned the horse?
- In what disciplines does he have training?
- Is he easy to catch?
- Is he registered?
- Is he suitable for a beginner rider?
- Is he up-to-date on vaccinations, dewormings, and farrier trimmings?
- Is he well-behaved for bridling, saddling, and mounting? Does he stand quietly?
- What breed?
- What is his age?
- What is his exact height?
- Why are you selling?

taken away from their herd companions so watch how he acts when he's by himself.

Watch the seller ride the horse before you or your companion attempt it. This is still a strange horse, despite how much information you may have obtained about him, and it's always wise to proceed with caution. Let the seller demonstrate the horse's ability at the walk, trot, and canter, going both ways in the arena. Observe the horse's general

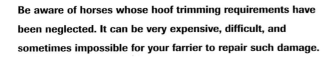

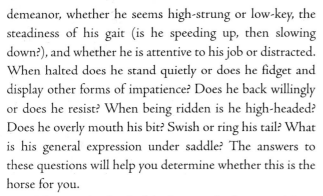

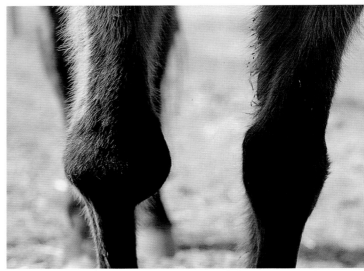

Be aware of horses whose hoof trimming requirements have been neglected. It can be very expensive, difficult, and sometimes impossible for your farrier to repair such damage.

Major blemishes that have the potential to compromise (or have already affected) the soundness of the horse should be a warning sign. You don't want to buy someone else's problems.

demeanor, whether he seems high-strung or low-key, the steadiness of his gait (is he speeding up, then slowing down?), and whether he is attentive to his job or distracted. When halted does he stand quietly or does he fidget and display other forms of impatience? Does he back willingly or does he resist? When being ridden is he high-headed? Does he overly mouth his bit? Swish or ring his tail? What is his general expression under saddle? The answers to these questions will help you determine whether this is the horse for you.

When you've finished looking at the horse and have asked any questions that you may have, it's perfectly acceptable to tell the seller that you're going to think things over and that you'll get back to him or her with your decision in a day or so. Obviously if the horse is completely unacceptable for your needs, you should tell the seller right away that the horse is not right for you. But if you do like the horse, there is no need to make a rush decision, even if you are certain that you'd like to buy it. Take the time to discuss the horse with your friend. Then, take a night to sleep on your decision and see how you feel the next day. If you decide to pursue the purchase, your next step is to finalize the details, which includes a pre-purchase physical exam.

PRE-PURCHASE EXAM

There are many reasons that you may want to have a pre-purchase exam (PPE) physical done on the horse you are considering. If your horse will be subjected to heavy training, if you're planning on competing in strenuous disciplines, and especially if the horse has already been competing or is in training, you will want to be reassured of his state of health. If you are purchasing the horse long-distance and won't be able to view the horse in person prior to purchase, it is especially important to have a veterinarian check the horse over for you. It is also a good idea to request a PPE if the horse is older. Advanced age means the horse has had more of an opportunity to have sustained an injury or he may have other health issues.

While it is always the best idea to schedule a PPE, some buyers will assume the risk of foregoing it, particularly if they are buying a young horse, a companion horse, or pasture pet; if they are not looking for a competitive show horse; or if they are purchasing locally and are able to view the horse in person. As with so many other purchase-related decisions, you have to make the choice with which you are the most comfortable.

If at all possible, you will want to arrange for the PPE to be conducted by a veterinarian who is not the horse's usual vet.

It can be an awkward situation if both the seller and the buyer are clients of the same vet because it can create a conflict of interest. Some vets will not perform the PPE if both parties are clients. Other vets will only do so with the stipulation that they are going to simply present the results, regardless of whose favor they may be in. If you must find a different vet to perform the PPE or if you're purchasing long-distance and need to locate a qualified veterinarian, you can ask for a referral from your local vet. He or she may be able to give you a good suggestion. Another option is to contact the American Association of Equine Practitioners for a recommendation.

Before the PPE begins, you'll want to talk with the veterinarian about your plans for your horse. A PPE for a competitive hunter will be different than a PPE on a yearling broodmare prospect, and your vet will need to know your plans in order to conduct the proper tests. Generally speaking, a PPE consists of a general, overall health examination including eyes, teeth, skin, heart, and lungs. This could also include jogging in straight lines and circles to evaluate soundness. Flexion tests on the legs can also be performed to test for any signs of lameness. The results of the flexion tests can help determine whether or not x-rays of the hocks, fetlocks, and feet are needed. Some PPEs include blood work for drug screening, as sometimes buyers like to be certain that the horse has not been drugged by the seller prior to the PPE. Unscrupulous sellers may do this to mask a lameness issue or temperament problems. In addition, if you're new to horses and unfamiliar with how to determine correct conformation, your vet can evaluate structural correctness during the PPE and give you an educated opinion.

After all of the tests have been performed and the exam is finished, you'll be able to discuss the results with the vet and then determine whether the horse is physically suitable for your needs. It's important to understand that the veterinarian can only give you his/her findings on that particular day. Obviously, the final choice rests in your hands, but it's always wise to consider advice when making such an important decision.

FINDING A HAULER

Let's say that you've found your dream horse on the internet and you watched the video and it far exceeded

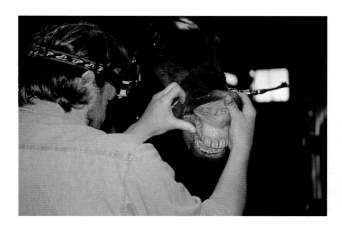

Your horse's pre-purchase exam should include a thorough examination of his teeth. This veterinarian is performing a check of this horse's teeth. This Thoroughbred gelding has his lip tattooed with his Jockey Club registration number, which can be checked as verification of his age.

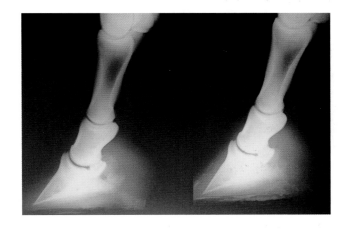

If you decide on investing in a veterinary pre-purchase exam, you may or may not elect to perform x-rays on the horse's legs. Horses that will be used in strenuous competition or in heavy training regimens should be x-rayed to detect potential defects, which may interfere with the horse's performance.

your expectations. You've shown the tape to half a dozen friends and they are all as excited as you are. You found a wonderful vet to complete the PPE and you had the x-rays taken as an insurance policy. The horse passed with flying colors and the vet proclaims him "healthy as a horse." You've sent your payment in full to the seller and the horse's registration papers are already being transferred into your name.

The ideal horse hauler will transport your horse from his current stable to yours in a timely fashion. Ideally the horse would be shipped directly door-to-door, but this option is not always possible and can be expensive.

There's only one glitch in this whole marvelous affair. The horse is in California, and you are in West Virginia. So, what's next?

Well, if you're like any enthusiastic horse buyer, the next step is to get your dream horse home. This could involve hooking up your truck and trailer and making a scenic trip across the country (see chapter 8 for the section on choosing a truck and trailer). Or you could hire a professional hauler.

To people who are unfamiliar with the horse industry, it often surprises them to hear that there are numerous companies devoted to hauling horses across the country from seller to buyer, mare owner to breeding farm, and show barn to horse show. These professional haulers make it their business to haul horses safely, affordably, and reliably. The question is where do you find one?

Again, as with horse shopping, the internet puts dozens of professional haulers at your fingertips, as many transport companies have websites. Some even have their upcoming schedules posted so that you can pinpoint if a hauler has a trip headed your direction. The cost of your haul can be less if the hauler is looking to fill a load that's already going from point A to point B. The cost will typically be higher if you want them to go to locations that are not on their schedule.

Begin by making a list of companies to contact. Besides your findings from the internet, you can locate many transport companies through advertisements in newsstand horse magazines. Ask for quotes, compare routes, and choose the best hauler you can find. Other horse owners may be able to give you recommendations on haulers, and you can also consult the seller for advice on the haulers he or she typically uses.

A more typical situation is where a hauler creates a route across the country or region and stops to pick up and unload horses several times along the way. Doing so maximizes his profits for the trip while still making an effort to move horses along quickly and offering customers a reduced rate.

There are a few things to keep in mind when booking a trip with a hauler. Ask about the hauler's hauling policies. Does he stop every few hours to allow the horses a chance to rest and drink? Does he keep hay in front of the horses during the entire trip? Will he be unloading at various points along the way, and if so, what sort of facilities will your horse be stabled at overnight during the trip? Will your horse be traveling loose in a box stall—believed by many to be the safest and least stressful form of transportation for a horse—or will he be tied for the entire trip in a slant stall? Will the hauler be making a direct trip from the pick-up point to the destination or will he be picking up and dropping off other horses (possibly in other states) along the way? It's a good idea to convey to the hauler that you'd like your horse delivered in as timely a fashion as possible. Your horse shouldn't be subjected to a longer trip than necessary and doesn't need a tour of the United States!

There is paperwork that is absolutely necessary to have prior to the horse being shipped, which include the following:

- Negative Coggins test result
- Health certificate
- Brand inspection (in states where applicable)

There may also be additional permits or forms needed for interstate travel, so be sure to check with your veterinarian or hauler, or you can contact the Department of Agriculture office for the states you will travel through regarding any other paperwork that you may need. For example, some states require that a negative Coggins test be current within six months, other states require it to be current within 12 months, and other states require that it be current within the calendar year. Make sure that you verify that your horse falls under state regulations.

LEASING

Some people feel more comfortable entering into a lease agreement before they make the commitment to purchase a horse. There are several reasons that people choose to lease a horse rather than buy. For the child rider who is going to be showing competitively, parents may opt to lease a pony that is the perfect fit for their child, then when the lease expires, move up to a larger pony or horse. This

During a lease, it is important for both parties to specify details, such as if the lessee will be able to show the horse and who gets to keep the ribbons!

prevents the need for a long-term financial commitment to one particular horse. You can also lease a more expensive horse than you could afford to purchase outright. In lease-to-buy agreements, you have the opportunity to lease a horse for a period of time, during which you can determine whether you and the horse are a good match. If everything is working out well, you can make the decision to purchase and the lease payments are applied toward the purchase price. However, if the match just doesn't seem right, you can part ways at the termination of the lease without further obligation.

There are a few things to keep in mind when pursuing a lease agreement. These will be detailed in the contract between the lessor (owner) and lessee (you). Make sure that the contract specifically details the exact terms of your agreement. In addition to the basic contract outlining the horse's description, the lessor and lessee, and the duration of the lease, you will want to make a notation regarding who will be liable if the horse is injured during the term of the lease. You may list provisions on whether or not insurance is required, as well as detail any usage limitations for the horse and who will have the final say on emergency care. In a typical lease agreement, the lessor retains the ownership of the horse and the lessee is responsible for expenses (board, feed, farrier), as well as for the daily keeping and care of the horse. It's always wise to have an attorney go over your contract before you sign it to ensure the contract is in compliance with state laws and regulations.

There are other leasing options available. A half-lease (also called a share-lease or a partial-lease) allows the lessor and lessee to split expenses and use of the horse. There are many ways to arrange the details of a half-lease, depending on the situation. Some half-leases involve a 50/50 split of all expenses with the figure fluctuating from month-to-month as expenses change. Other half-leases involve a specific monthly fee which the lessee pays the lessor, regardless of exact expenses for that particular month. Many people find this to be the easiest, as both parties know ahead of time what the fee will be each month and eliminates any chance for surprise or misunderstanding. Another factor that you will need to determine is who will have access to use the horse and when. Again, this will depend entirely upon each unique situation. For instance, the lessor may have lessons at the boarding facility on Mondays and Thursdays, and the lessee may have lessons on Wednesdays, Fridays, and Saturdays. Therefore, the contract would stipulate that each party will have access to

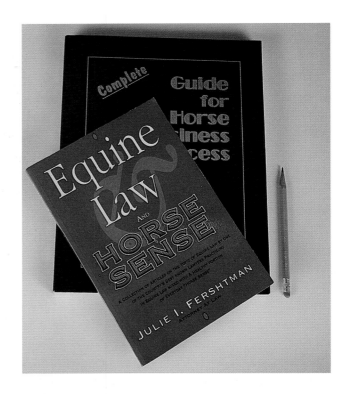

Researching contracts may be important to make sure you understand all details involved. However, keeping things simple is probably all that is needed on a basic purchase or lease agreement.

If the seller permits it, taking a horse home to try him out in the environment of your stable and grounds is a great way to get a feel for him and his fit on your farm.

Make sure you ride the horse in the way that you plan to use him if you were to purchase him. See that he willingly performs the things you will require of him.

the use of the horse on those specific days, which eliminates any chance of both parties wanting to ride the horse on the same day. It's also necessary to specify who will have use of the horse for showing purposes.

There are also other points that your half-lease contract might include, such as standard of care (grooming, care of tack, blanketing); who feeds and when; where and under what circumstances the horse can be ridden; who is permitted to ride the horse (lessee's friends or family); who is responsible for veterinary care; whether trail riding is permitted; whether treats are allowed, and so on. It's possible that the lessee will need to sign a release form stating that he or she will not hold the lessor responsible for any injuries incurred during the use of the lessor's horse. In addition, the lessee may be responsible for injury to the horse due to his or her own negligence. Considering all of these points ahead of time can help make your leasing experience a smooth one. Many people have been very satisfied with their half-lease agreements and heartily recommend them.

Another important point to remember is that breed organizations often have regulations on leasing and whether or not it is allowed. Some breeds will not permit leased horses to compete in sanctioned breed shows, while others allow it as long as the lease is officially recorded with the association.

TRYING A HORSE

When you're trying to decide if a horse is a good match for you, people may tell you to see if the seller will consent to a trial period with the horse. When you take a horse on trial, you are allowed to take the horse home to your barn or boarding facility and spend a period of time evaluating how well you and the horse suit each other. If you like the horse after the trial period (typically one to two weeks), you pay the seller and purchase the horse. If you don't like the horse and don't feel that you are a good match for each other, you return the horse to the seller. In theory, this scenario works and prevents buyers from making purchase mistakes and gives sellers an added comfort of knowing that their horse is going to a home that is pleased with their new horse.

Unfortunately for buyers, many sellers will not allow a trial period for several very legitimate reasons. Unscrupulous buyers will sometimes take a horse on trial when they are not seriously interested in purchasing it, effectively borrowing and obtaining full use of a horse without cost. Sellers sometimes receive horses back after trial periods in considerably worse condition. Many sellers are reluctant to hand over all responsibility and care of their horse to a stranger whose standards of horse care may or may not be acceptable.

Congratulations! You've found the right horse that you and your family are happy with and one that will provide years of fun and excitement.

If a seller is willing to allow you to take his or her horse on trial, you may be asked to place a non-refundable deposit on the horse or leave a check for the entire purchase price (to be cashed if the horse is not returned in the same condition after a specified period of time), or the seller may ask you to insure the horse for the duration of the trial. In any case, a contract detailing the specifics of your agreement should be signed by both parties.

But what if the seller's policy is not to allow a trial and all of your horse friends feel that you shouldn't purchase a horse without one? What's a buyer to do? First of all, don't assume that the seller is hiding something because he or she doesn't allow trials. Many perfectly reputable sellers have a policy of not sending horses on trials. It may have nothing to do with the horse, and it certainly doesn't have anything to do with you personally. Find out whether you can visit the seller's farm on several occasions to try the horse under a variety of circumstances (different times of the day, different routines, etc.), see if the seller will haul the horse to a different location (perhaps another trainer's barn), or see if you can ride the horse at a show. These options will allow you to evaluate the horse in other environments than simply riding around his or her home arena. If you have reason to suspect that the horse has been drugged to make him quieter while you ride him at the

seller's barn, you can have him tested for drugs to set your mind at ease.

Some sellers will offer a return policy, which is essentially the same as a trial period except that the seller is paid in full ahead of time. If after a set period you discover that the horse is not suitable for you, the horse can be returned in the same condition (some sellers stipulate "physically and mentally") for a refund.

CONTRACTS

Whether you're buying or leasing, it's important to make sure that you and the seller sign a contract outlining the details of your agreement. While contracts often vary in length and content, there are certain basic elements that need to be included before both parties sign. Some sellers have a standard purchase agreement or contract that they use for every transaction, while others may tailor their contract to suit each unique situation. If you're purchasing from a backyard horse owner, he or she may not have a standard written agreement prepared, so you may need to come up with a contract yourself.

The basics of your contract should include: identification of parties (seller and buyer); description of the horse including color, markings, age, sex, registration number (if applicable), sire, and dam; purchase price; terms; and

Shows, trail rides, peaceful jogs around the arena, and more await you when you finally get your new horse home and begin the enjoyable lifestyle of horse ownership.

warranties (if applicable). If everything is spelled out in black and white and signed by buyer and seller, there can be no questions later regarding the specifics of the purchase. Although it's certainly less complicated if you're purchasing a horse and paying the entire purchase price at once, the situation becomes considerably more involved if you're purchasing on terms or leasing. In those instances, the contract should contain clauses that define which party is responsible for routine items such as feed, farrier, deworming, and vaccinations, as well as unexpected expenses such as veterinary care. Be certain that you clearly specify the financial arrangements, how much is due per month, when payments are due, at what point registration papers are transferred, and whether insurance is required. There are excellent books devoted to this subject and many contain sample contracts you can tailor to your own needs. By using their models, you can help ensure you aren't forgetting a vital part of the contract. In the event of a dispute, your signed contract will help protect you, whereas a verbal agreement and a handshake will not. It's a good idea to have a lawyer look over your contract to verify that it is in compliance with state laws and regulations.

INSURANCE

As we have discussed previously, there are several reasons that you may want to consider insuring your horse. Buying a horse can be a major purchase and expenditure. Taking out a mortality insurance policy can protect your investment in the case of death, whether it's due to illness, injury, accident, or theft. In addition, a major medical policy can protect against unexpected veterinary expenses, such as surgery. If you're leasing a horse you may be required to insure him to prevent any financial loss to the owner. Similarly, if you're purchasing a horse on terms and the horse is staying with the seller until final payment is made, the seller may insist upon insurance to protect all parties against loss.

There are many insurance companies to choose from, and you'll want to take some time to compare their policies to make sure that you choose the best one for your needs. Some have higher deductibles, some will pay more than others for surgery, some companies require yearly examinations prior to the coverage continuing, and some companies will not insure a horse with a pre-existing medical condition. Other companies may not insure a horse who has had colic surgery, so make sure that you ask suitable questions prior to choosing a policy.

GETTING TO KNOW YOUR HORSE

This horse exhibits a slightly interested but reserved expression. The horse is watching something, possibly his owner or another animal who has aroused his curiosity.

By now, your horse is home and getting settled in and you now have the opportunity to get to know him more thoroughly. Certainly, you got a glimpse of his personality and character while you were shopping, but now you will have the opportunity to learn all of his little habits, quirks, and routines face to face every day. Don't worry, you'll have fun!

CHARACTER

Like people, horses come with unique and interesting personalities. You'll have a good time learning about your horse's personality characteristics. It's one of the very special parts of horse ownership.

You'll begin learning about your horse's character right away by watching how he relates to you as you work around him. Watch his expressions as you bring his food, clean his stall, or refill his water bucket. Does he stay away but still view you with interest, eyes brimming with curiosity at the task you are performing? Or perhaps he moves to the opposite corner of the stall, face turned into the corner, as if pretending that you aren't there. He might follow you around, sniffing your pockets, trying to chew on your stall pick, and begging for attention.

In the first instance your horse is curious but cautious. In the second instance your horse is shy and nervous. In the

third instance your horse is bold and brave. Bold horses are the ones who seem unafraid of anything, are unflappable in new situations, and are always the first to greet a stranger or check out an unfamiliar sight. They don't bat an eye when you lead them past something new, don't care if you switch them from one stall to another, and enjoy checking out new situations. These are the leaders in pasture situations. They are the horses who will stand at the fence line when a new person arrives at the barn, will be in your way the entire time you try to repair a fence, and will investigate anything unusual or interesting. They are inquisitive, curious, and ready for anything.

Nervous horses are the absolute opposite. Simple objects can scare them, new situations make them upset, and they shy away from strangers or new sights. Walking past a bag of shavings that is in the wrong place can be extremely upsetting. They despise being moved to different stalls or turnout locations and tend to be the horses that are particularly attached to their daily routines. So if you have the patience to work them through their nervous worries and convince them that everything is all right, then you could end up with a very willing and responsive horse with whom you have an excellent camaraderie.

The curious but cautious horses are the type that like to think things over first, but are completely happy once they decide that everything is all right. The key to working with and handling them is to let them have a moment to process the situation before proceeding. Giving them a moment to think can make new situations and sights much easier because they really are curious and want to know what's going on. Kind words and gentle handling can really make a difference with this type of horse. A word of encouragement can work wonders in a horse who is trying to decide whether something is scary or not. Deep down he really is interested and curious about new things, but it just takes him a moment to let his curiosity overcome his momentary fear.

When you bring any new horse home, it's important to remember that you must give him time to settle in. Often in our excitement and haste to get going with our new horse, we forget the complete upheaval that the horse has just gone through; removal from a familiar place, a trailer ride (sometimes taking several days), a host of strange humans, a new barn, new horses, new sights, new smells, and new routines. The horse is attempting to understand and process all of these new things, not to mention trying to establish himself within this new situation and quite possibly find his place among new pasture mates. Is it any surprise that horses sometimes act up unexpectedly when transported to their new homes? Horses need time to adapt to their new situations, become friends with their new people, and rationalize the sights and sounds of their new surroundings. In some horses this adaptation can be done in a few minutes. One good look around the place and they are happy. In other horses this may take several days or a couple of weeks before they are fully adapted and relaxed in the new atmosphere. For certain horses, specifically youngsters, this change could literally take a few

Some horses are bold, curious, and will explore any new object without fear. The curiosity of their surroundings and interest in new things will help them overcome any anxiety.

months. This is especially true for young horses that have been transported for long distances at young ages. While most settle in without much fuss, there are some individuals who are somewhat traumatized by their experiences and it takes them a longer period of time to recover from the mental and emotional stress.

The bottom line is if you've just purchased a horse and you feel that he's not behaving the way you expected him to, try giving him a little more time to settle in before you write him off as unreasonable or bad. Spend as much time as you can giving him positive attention, such as a lot of scratches on the neck, quiet grooming sessions, and plenty

Horses have a wide range of vocal expressions they use in various ways. Among others, there are lonely whinnies used when they are left alone in the pasture without a buddy, and "feed me" calls for when they hear the grain bucket coming.

of kind words. Your days of horse shows and trail rides will come a lot faster if you've given your horse some time to adjust before you press him to perform.

Horses have exceptional memories and you'll be amazed by their display of understanding. One instance from our farm was particularly memorable. We weaned a filly from her dam and turned the filly out with some other young horses where she spent a year at pasture, grazing and growing up with the others. In the meantime, her dam raised another foal, which was subsequently weaned and sold across the country. At that point, we had to do some rearranging of pasture mates and we decided to turn the filly back out with her dam. There was a look of joy on the filly's face when she recognized her mother after all those months. With an excited whinny and a fast gallop across the field to see her dam, the filly promptly put her head underneath the mare's stomach, as if to nurse. Well, of course her dam

would have none of that and the filly didn't try it again. The two of them took off across the pasture and in moments were happily grazing side-by-side. After long months of separation, the filly had still not forgotten her mother.

Another memorable occasion from our farm involves an article of clothing. Here in the frozen north country of the United States you are never far from your heavy winter jacket from November through April. One of our mares has associated the sight of a particular blue jacket as being the "sugar coat." This older broodmare will come up with the most animated expression whenever she sees the jacket and proceed to sniff each and every pocket. The expression on her face is priceless: "Where on earth did you put the sugar? It was in the blue coat one other time!" In this instance, the sight of the specific coat is the mental stimulus that triggers her memory. She never sniffs the pockets of any other jacket at any other time.

Foals start to whinny early on, often within the first hour of birth. Foals learn to follow the examples of other horses in their environment and will call out if a friend leaves the field or if they wander too far from mom.

These two horses have their necks arched and are sniffing each other's noses. They are strangers and are sizing each other up to determine where they belong in the pecking order.

Both ears pricked forward is a sign of an interested horse who is trying to determine what he is seeing or hearing, even though he may not be particularly afraid of it.

With one ear forward and the other back, the horse is mildly interested in his surroundings but peacefully waits for something more interesting to happen. When being ridden, a horse will often flick one ear backward like this to listen to what his rider is saying to him.

COMMUNICATING

If you were like any self-respecting horse-crazy child, you probably spent a fair portion of your childhood reading fictional stories about horses. We could safely wager that you probably read your fair share of *The Black Stallion, The Golden Stallion,* everything written by Marguerite Henry and Dorothy Lyons, and anything illustrated by Sam Savitt. You probably read plenty of books where the protagonist had long conversations with her horse and the horse seemingly understood everything the girl said. And if it was an illustrated book, chances are the horse answered

back! Now maybe your horse isn't quite as vocal as that, but he does have his own ways of communicating if you're watching closely enough to notice.

Obviously whinnies, nickers, and squeals are the audible forms of horse communication. There are few things as rewarding as the chorus of early morning whinnies as you enter the barn, and it's certainly heartwarming when your horse looks up at you across the field and nickers. If you have more than one horse, it won't be long before you can easily distinguish between their voices.

Both ears sideways or slightly back is an indication that the horse is completely peaceful and relaxed and not concerned at all about what is going on around him, probably because nothing is!

There are nearly 20 horses at our facility and we can usually pinpoint each horse's whinny with fairly certain accuracy. Each one is distinct and unique; some are high, some are low, some scream, and some neigh. Interestingly enough, the specific types of whinnies seem somewhat genetic. We have three generations of a particular line of Welsh Ponies who have the most bellowing type of whinny. They are the only ones we have that have a distinctive hollering sound, but it's always cause for a chuckle when the latest foal from the line is born. The foal lays there in the straw, soaking wet, and lifts a wobbly head into the air, letting out that distinctive bellowing whinny, and we say to each other, "Yes, he certainly belongs in the family!"

A horse that rubs its chin on your arm is quietly communicating its need for a scratch on an itchy spot. A horse that nuzzles its nose on your pocket is (you guessed it!) asking for a treat. A horse that's stomping, swishing, covered in bugs, and standing at the paddock gate is asking to be brought into the barn.

It's very important for you to communicate as well. Talking to your horse helps to relax him, reminds him where you are so that you don't startle him, and makes the atmosphere much more pleasant than dead silence. Frequent repetition of specific words, such as "whoa," "no," "quit," "back up," or "good boy," when accompanied by the appropriate gestures, helps your horse understand and learn the proper response when you state your command. You should also regularly call your horse by name. Horses are intelligent individuals and most will easily learn their names. We once had a yearling filly who would whinny in reply whenever we called her name.

WHAT'S HE SAYING?

Horses communicate feelings and emotions through a broad range of mannerisms. Some are obvious and others are barely detectable. Here is a sample of the most common equine behaviors and what they can mean.

Ears pinned back is a sign this horse is not pleased about something and may be giving a warning to other horses (and people) around him, saying, "Watch out."

Biting (other horse)	Establishing pecking order, warning, anger
Biting sides	Irritation, possible colic
Blowing loudly, tail up	Excitement, exhilaration, possibly fear
Bucking (at liberty)	Playfulness, frolicking
Ears pinned back	Anger, discomfort, warning
Ears pricked forward	Interested, attentive, focused on a particular sound or object
Ears sideways	Boredom, resting, relaxed
Laying down	Resting
Laying down and getting up repeatedly	Possible colic, pain, attempting to escape from flies
Neck arched, teeth clicking	Young horse or foal; submissive
One ear flickering in a direction	Listening without looking
Pawing ground	Irritation, impatience, discomfort, or about to roll
Rear leg cocked	Relaxation, resting
Rear leg stomping	Annoyance (flies)
Rearing (at liberty)	Playfulness
Rolling	Pleasure of being turned out, enjoyment
Rolling multiple times	Possible colic, pain, attempting to escape from flies
Striking front leg out, squealing	Response to meeting unknown horse
Swishing tail	Annoyance, aggravation
Teeth grinding	Anxiety, stress
Two horses biting each other's withers	Mutual grooming: bonding, establishing relationship
Upper lip raised	Flehmen response: reaction to strange odor

SAFE HANDLING OF YOUR HORSE

Learning to properly handle your horse can help prevent accidents and potential injury to you or your horse. Your number one concern at all times should be safety, and it's important to always be fully alert, pay attention, and keep a watchful eye so you can help avoid a potentially dangerous situation. If you make it a habit to always be safety conscious, it will become second nature to you and you'll always be unconsciously avoiding potential problems or situations.

Simply knowing your horse's personality and predicting his responses can help you avoid trouble. If you know that he is afraid of the riding lawn mower, you can avoid leading him near it while someone is using it.

PROPER AND IMPROPER WAYS TO LEAD

Leading a horse is easy, just put on the halter, grab the lead rope, and off you go. Nothing to it. That is, there's nothing to it if you do it correctly. Leading a horse improperly may potentially set you up for a dangerous situation, so it's impor-

Catching Your Hard-to-Catch Horse

Your horse has been turned out for the day and you're ready to bring him back to the barn for the evening. Armed with your lead rope and halter, you head out to catch your horse. If you're fortunate, your horse will be glad to see you and eager to come inside. If you're not so fortunate, he might take one look at you and run.

For a horse who needs a bit of encouragement, it may be a good idea to bring along a bucket of feed to entice him to be caught. The key, as illustrated in this photo sequence, is to remain calm and collected. If you get frustrated or impatient, your chances of catching your horse are quickly reduced. Approach your horse slowly with kind words and sweet tones. Offer your horse the reward of a treat and quietly slip on the halter. If you are able to catch him quickly without a dramatic event, he will be less likely to consider it a game that you two play and more likely to come in and be caught without any fuss. In any case, if there are multiple horses in the field, bringing grain out to the group can cause more problems, including the possibility of you being mobbed. Therefore, grain should only be used in the situation of a single horse who is resistant to being caught.

Foals and older horses love to get down and roll, especially if they have been inside a stall and were just turned out to pasture.

Two horses who are chewing or biting on each other's withers are establishing a relationship and friendship, as well as enjoying the pleasant scratching.

tant to know how to do it correctly so that you and your horse can be safe.

To lead a horse, you always stand on the left side and hold the lead rope in your left hand. The lead rope should not be left long to drag on the ground. It should be folded neatly in your hand. Never wind the lead rope into a circle and hold it because if the horse were to bolt or spook, the circle could tighten around your hand and you would be unable to escape.

In your right hand, grasp the other end of the lead rope; the part just underneath the buckle that attaches to your

Some horses love to rear when they are playing and wouldn't think of doing it at any other time. Fresh, windy days seem to make all horses want to get out and play hard.

With a rear leg cocked, this horse is relaxed and resting while he is tied to his trailer, waiting for his owner to return and perhaps tack him up for the next class in a horse show.

When a horse curls or raises his upper lip, he is displaying the Flehmen response. This is a reaction to a strange odor, although it is most commonly seen in stallions who smell signs of a mare.

horse's halter. Walk along at your horse's shoulder, not too far in front so that you're dragging him along, but not so far back that he is dragging you. You are the one holding the lead rope and you should have confidence and an authoritative air. Otherwise your horse may get the idea that he is the one in charge and that you are merely along for the ride.

WORKING AROUND YOUR HORSE

Safety should always be foremost in your mind as you work around your horse. Be vigilant about keeping yourself out of harm's way and never assume that your horse won't do something unexpected, even if he's old and unexcitable.

If you want to tie your horse while you work around him, make sure that you always use a quick release knot so you can quickly untie him in the event of a problem. You should never tie your horse to an insecure object, such as flimsy fencing or a weak post. Horses are incredibly strong

and when they throw a half ton of body weight against something, it needs to be very strong to withstand the strain. Never tie your horse with the reins from his bridle, only use a halter and lead rope for tying up the horse.

Don't sit down next to your horse to groom his legs or clip him. Always stay in a squatting position so if he shifts around or moves, you'll be able to quickly move out of the way.

YOUR HORSE'S ROUTINE

You have probably established a routine of sorts, meaning that you feed your horse at the same time each day and have regularly scheduled times for turnout and stall time. Your horse also has his own routine, and by paying attention to it you'll begin to understand more about his personality and what is (or isn't) normal for him.

Some horses are very predictable in their habits and by noting their daily patterns, you can either rest easy

Horses biting each other are either playing or establishing a herd order. Young colts and fillies will play-bite each other just for the fun of it.

knowing that they are healthy and happy or if they aren't acting normally.

We have one mare who will drink three gallons of water after being fed in the morning. There are others who don't drink a drop after they eat. Neither procedure is preferable, they're just different.

There is another mare here who will eat her evening hay and go all night without drinking any water, but then she will drink half a bucket while she munches on her morning hay. We could be concerned when we go out to the barn first thing in the morning and see that she hasn't

touched her water all night, but knowing her routine, we know that when we come back in an hour, after morning hay, she will have taken a good drink.

Some horses will eat any morsel of food that is available to them and would never dream of leaving even a wisp of hay in their stall after a meal. Other horses are more nonchalant about their food and will regularly leave a portion of their hay. Now, you could worry that something is wrong with them because they haven't finished eating. But knowing the personalities of my horses helps me understand that particular ones are just less compulsive

This horse is bored with his current situation and is tired of waiting at his trailer. He informs the world of this information by pawing the ground.

This mare is telling her nearby pasture mates to stay clear and that she is the one in charge. Kicking is used occasionally by members of the herd to maintain the dominance.

Always lead from the left side of the horse, between his head and shoulders, to give you precise control of his head and body while keeping you out of the way of the lead rope.

A commonly seen mistake is a person leading his or her horse from far out in front with the lead rope tangled. Don't do this even if you are leading an old gelding who never spooks. This situation puts you in a vulnerable position and you have no way to get control of the animal quickly if necessary.

about eating everything in sight and some like leaving hay in the stall, even when they are perfectly healthy and feeling great.

Your horse will usually nap at regular times during the day, and have other times where he is more restless or playful. If your horse is used to being turned out to pasture every morning at 7 a.m., he may be uncooperative and spirited if you decide to saddle him up for a ride at 7 a.m. This

is not because he is naughty, it's because his routine has been disrupted, and he's upset and confused. The better idea would be to turn him out early (say at 6 a.m.), give him an hour or so of play time so that mentally he has had the refreshment of turnout, and then catch him up and go for a ride. You may find that you have a much more willing and cooperative partner, and all because you gave a little extra thought to making the routine change a little easier.

FEEDING YOUR HORSE

Horses are happiest when left to graze for hours with their pasture buddies. This situation resembles the natural habits of wild horses and offers many health benefits for your horse.

Feeding a horse sounds easy. After all, you can turn him out on pasture and he will eat all the grass he wants. If you don't have access to pasture, you can throw him some hay once in a while. While these ideas may sound plausible, they underestimate the vital importance of feed and nutrition to the health and well-being of your horse.

So, does that mean that feeding a horse isn't easy? Will all your future days be filled with analyzing supplements, grain rations, and weighing hay? How will you know if your feeding program is adequate? How will you know if your horse is getting all of the essentials necessary to a balanced diet? How much food is too much? How often is too often to feed? Should you use round bales or square bales? These questions can seem overwhelming and daunting, especially when you're trying your hardest to provide a healthy and well-balanced feeding program for your horse.

Even though planning a feeding program isn't as simple as you might think, it isn't rocket science. With some research and study, you will be able to feel confident in your feeding choices and satisfied with your horse's condition and nutritional health.

BASIC NUTRITION

Horses are grazing animals and are happiest and healthiest when they have access to a continuous source of food and water. The horse's stomach is rather small and not designed to store large amounts of food, hence the horse's natural desire to graze continuously, thereby keeping a small amount of food in the stomach at all times. Obviously, not all horses are—or can be—kept in a 24/7 grazing situation so you should give careful consideration to your feeding program in order to give your horse the healthiest and most nutritious diet possible, particularly if he will not have access to a pasture around the clock.

The bulk of a horse's diet should be made up of roughage (grass or hay). If this roughage does not come from grazing, then you will need to replace the grass intake with hay. Many horses who are not working or regularly exercised will be able to maintain a healthy weight simply through their intake of roughage. Horses who are working will benefit from the addition of concentrates (grain) to their diet, with the amount varying by circumstance. A well-balanced equine diet includes protein, carbohydrates, water, minerals, and vitamins.

PROTEIN

Horses receive a good portion of their protein intake through grass or hay, but they also occasionally obtain protein through additional carbohydrates will be necessary if he is frequently worked. These additional carbohydrates can come from grain. It is important to provide enough energy from carbohydrates so that the horse's body is not forced to convert protein to energy, thus depriving the body of enough protein to maintain healthy growth.

WATER

Water is as important to a horse's health as feed. Horses must have continuous access to fresh, clean water to ensure optimum health and well-being. The average horse will drink 1 gallon for each 100 pounds of body weight per day. Obviously, this amount will vary within individuals, especially dependent upon the weather and temperature extremes, but it is a good rule of thumb. Historically, many horses have been managed successfully even when water has been provided only at scheduled intervals two or three times a day. However, it is certainly more beneficial to the horse to have constant access to water and the ability to drink whenever he is thirsty. Dehydration can cause lethargy, weight loss, and an increased risk of impaction colic.

Water should be clean and fresh, which means that you need to be scrupulous about maintaining clean water buckets. Wash water buckets frequently to prevent the growth of algae and remove particles of food. Some horses are reluctant to drink stale water and will drink greater quantities when the water is frequently refreshed. Special care will need to be taken during the winter months when the water buckets and troughs will be very cold or frozen. Frigid water is very unpalatable for horses and incidences of colic tend to be higher in the winter from decreased water intake due to cold water. Heated water buckets or water heaters for troughs can be very beneficial and make horses more likely to drink. If heated water isn't an option, you will want to remove the frozen water and provide fresh water at intervals throughout the day so that your horse isn't going without water for hours on end.

Providing salt blocks is a good way to increase water consumption at any time of the year. If you suspect that your horse is not drinking as much as you would like, you can also try adding loose salt to his grain, which can help to increase thirst and water intake.

Horses who are kept in non-pasture situations, such as this, will require a considerable amount of hay to make up for the lack of available grass.

their consumption of grain. Protein is vital to the body's building blocks of muscle, bone, and skin. Too little protein in the diet can cause stunted growth and other problems, but too much protein can be a strain upon the kidneys, therefore care should be taken to avoid either extreme.

CARBOHYDRATES

Carbohydrates provide energy and much of a horse's carbohydrate consumption will come from grass and hay. While the amount of carbohydrates from forage is enough to maintain a horse's condition when he is merely existing (i.e., not in a regular training or working program),

MINERALS

There are many minerals necessary to your horse's diet but the vast majority are only needed in trace amounts. They are probably already provided for in your horse's general diet, so supplements are usually unnecessary. The exception to this generalization is salt, which needs to be provided for your horse's consumption. You can feed salt in block form or in loose form in a feeder. In either case your horse should be able to have access to as much salt as he desires. An increased salt intake usually leads to increased water intake.

VITAMINS

As with any other animal, vitamins are a necessary part of a horse's balanced diet. However, it is important to remember that vitamins are typically provided in ample amounts in a horse's general diet of hay, grass, and grain. Additional supplementation is generally not necessary, and overdoses of vitamins can cause more harm than good.

FEEDING

Now that you have an idea of basic nutrition, you need to implement this knowledge into a feeding program. How do you determine which feeds are right for your horse, what type of hay to use, how much to feed him, and how often?

MEASURING FEED BY WEIGHT

It's important to note that you should always measure feed by weight and not volume. It is much more consistent to feed three pounds of grain at each feeding than it is to feed three scoops of grain. The same is true for flakes of hay, as the variable size of the flakes can lead to overfeeding at one meal and underfeeding at the next meal. If you weigh the amount prior to feeding, you will ensure that you are giving your horse the same amount each time.

REGULAR FEEDING TIMES

Your horse will quickly become accustomed to your feeding routine and will learn to expect and anticipate when meals will occur. It's important to be as regular and reliable as you can with regard to the timing of meals. You don't want your

Types of Feed

There are many different types of concentrates, and the variety of choices can seem a little overwhelming. The type(s) of grain you feed will be dependent upon many factors.

OATS

The high energy content of oats makes them a popular choice for horses who are being heavily worked. Oats might not be the best choice of feeds for a horse who is not using up the energy by working, as it may increase their boisterousness (hence the phrase "feeling his oats"). You can purchase oats in various forms, including whole, rolled, bruised, or crushed. Some experts feel that whole oats are not digested as well as the other forms, but this is not a proven theory and many individuals have achieved satisfactory results by feeding whole oats.

BRAN

Wheat bran is typically soaked in water and served as a bran mash. While not considered an everyday feed for the majority of horses, it is sometimes fed to geriatric horses who have slower digestive systems because it is valued for its laxative effect, though that fact is disputed by some. Bran mashes are often given to mares after foaling to help prevent constipation after delivery. Some horse owners will give a bran mash after their horse has finished working for the day.

PELLETS

Nowadays many horse owners prefer pelleted feeds, which are commercially manufactured mixtures of concentrates in pelleted form. This type of feed offers convenience and reliability. While not a traditional feed, pellets are more thoroughly digested than single grains, and there is generally less waste. The uniformity of the feed is another benefit and the pellets are easily measured to ensure that you're able to feed the same amount each day. There are various forms of pelleted feeds. Some are simply mixtures of grains, while others are what is known as complete feeds and contain hay as well as grain. These complete feeds are designed to provide all of the nutrients necessary for a horse so that no additional hay or grain is needed. This can be ideal for an aged horse that may have trouble chewing and digesting food. However, many owners prefer to feed an amount of roughage, in addition to the complete feed, because horses are designed to eat throughout the day, and they crave the chewing motions of grazing or eating hay. The additional roughage also provides the horse with added bulk in his stomach.

SWEET FEED

The more traditional mixed feed, sweet feed is sold under many brand names and is often available from local feed mills who prepare their own mixes. Essentially, sweet feed is a mixture of oats and molasses. Other grains, such as corn, are sometimes included. The added molasses makes the feed more palatable to some horses who are picky about their feed. Sweet feed causes concern for some horse owners, as the higher sugar content from the molasses can be undesirable for horses with insulin resistance, Cushing's disease, or those who have previously foundered.

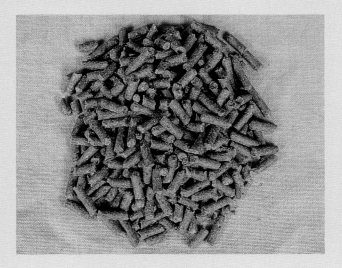

A horse's exercise level may impact his nutritional needs. Naturally active horses or those in an exercise program may require a more elaborate feeding program to compensate for the additional use of energy.

horse becoming stressed because you are three hours late with his morning grain. Horses are creatures of habit with strong internal awareness of time and schedule and it is your responsibility to set a schedule and do your best to stick to it. It's not uncommon for a horse to spend his day happily grazing at pasture, only to look up at the appointed time (and you'll be surprised at how accurate horses are!) and proceed to the gate, knowing quite well that grain time is imminent and you will soon be on your way with the feed.

Unless your horse has access to around-the-clock free choice hay or is out 24/7 on good pasture, you'll want to offer hay at intervals throughout the day, including a good breakfast in the morning and a large feeding of hay in the evening. Ideally horses should be fed three times a day with

a feeding of hay sometime during the early afternoon. If you are also feeding concentrates it is preferable to feed the grain more than once a day as well, especially if your horse is getting large amounts (over eight pounds per day). Performance horses in rigorous training often need to be fed very large amounts of concentrates and in these extreme cases, the feedings should be divided into three feedings or more. Smaller meals are easier for a horse to digest, and feeding smaller amounts more often helps keep a more continuous flow of food moving through your horse's digestive system.

CHANGES IN DIET

It's important to remember that you should always make any changes to your horse's diet slowly and ideally over the period of several days to a week. Abrupt changes in feeding schedules or amounts can cause a horse to colic and should always be avoided. This applies to any type of feed, including grain/concentrates, hay, or grass. If you're changing to a different type of concentrates (or a different brand of pelleted feed), introduce the new feed very slowly by mixing a small amount (perhaps 1/3 of the ration) in with the feed that your horse is used to eating. Many horses are suspicious of anything new in their feed buckets and will refuse to eat anything that doesn't smell "right." If you make the new ration small and mix it in well with the feed that your horse is used to, you can gradually get him used to the new feed without any upset of digestion or suspicion on your horse's part. Similarly, it isn't wise to abruptly switch from one type of hay to another, particularly if you are changing to a richer hay, such as a change from timothy to alfalfa. Take your time and begin by substituting 1/4 of the old hay with the new hay, and feeding 3/4 old hay. Do this for a couple of days, then alter the ration to half and half of each kind of hay for another two days. Follow this up with 3/4 new hay and 1/4 old hay for two more days until you've finally reached 100 percent new hay by the time a week has passed.

If your horse is stressed for any reason, such as being trailered, it is wise to decrease his amount of grain for a few days until he has settled in and relaxed. Even if your horse is afflicted from momentary stress, such as being hot after being heavily exercised, it's a good idea to cut back on his grain ration for a feeding or two. This prevents your horse from having a large meal of grain when his digestion might not be optimal.

TREATS

If you're like most horse owners, you won't be able to resist the thought of bringing along a treat for your horse every time you head out to the barn. Many people recall their childhood days when they would feed their pony any and every type of food, from pizza to cookies to bread and a little soda on the side to wash it down! While this makes for fond memories, it's probably not the wisest choice for your horse.

A good treat for your horse can be as simple as carrying

Always strive to hang water buckets in safe positions off the ground and away from where they could be tipped or tangled in. Buckets hung above the ground will also naturally stay cleaner and require less maintenance from you.

The regular washing and refilling of water containers is a simple chore that does not take very long but will greatly discourage the build-up of algae and maintain a clean water supply for your horse.

Horses love to play with water buckets and troughs when they're bored. Try to eliminate this situation by fastening the containers securely. It might be a fun game to him, but you don't want the horse stranded for extended periods of time without a drink.

a handful of grain in your pocket as a reward for being caught or just as a treat when you pass by his stall. You can try sugar cubes, which the majority of equines will enjoy. Carrots and apples are also perennial favorites, but you must be careful not to feed these in small chunks, which could pose a choking hazard. There are also commercially

Multiple feeding times may become increasingly important during the winter months when the horse will require additional roughage to make up for loss of energy from staying warm. What better way to start your day on a chilly winter morning than to be out in the crisp air feeding horses during a glorious sunrise?

Your horse will certainly be waiting for you if he knows he will be fed at a particular time. Horses love to figure out the predictability of their feeding schedule.

sold horse treats, such as apple-flavored biscuits.

Some people discourage the feeding of treats to horses, usually for the valid reason that feeding treats can often cause horses to become nippy and pushy. If you feed your horse a sugar cube every afternoon before you turn her out, she will become accustomed to this charming little routine and come to expect it. Thus, when you show up at the gate, she may search all of your pockets, bump you in the stomach with her head, or even bite your hand in search of the treat. On the other hand, there are some horses that can be given treats every day and never expect it, are always surprised to see it, and would never dream of nipping or pushing. These not-so-clever souls are always a joy to have around the barn!

If any signs of nipping or pushiness arise you should not feed that particular horse treats at all. It is not worth the risk to you. However, if you enjoy feeding treats and

It is very important to store your grain in a cool, dry container safe from insects and rodents. This stable manager has accomplished all of these goals in an ideal situation. You will need to incorporate these values into your own storage system.

A surprise treat once in awhile can help establish a quality relationship with your horse. He will learn to associate your companionship with positive moments, not always with ones he may dislike, such as being caught.

your horse enjoys receiving them, the best advice is not to eliminate the practice entirely but to offer treats occasionally and not every single day. Vary the times that you give them so that your horse cannot come to anticipate it at a particular time or place. Spontaneity is the key to successful treat feeding.

PASTURE

Good quality pasture is a valuable asset to any feeding program. Generally, horses are happiest when they are grazing. Horses with access to pasture typically display fewer stable vices (cribbing, weaving, etc.) than horses who are housed in stalls for much or all of the time. Many horse owners provide a compromise and offer turnout for their horses on grassy pastures during the daylight hours and putting them in their stalls at night. Or some owners, particularly those in the hot, humid south, will stall their horses during the day and turn them out to pasture at night when the temperatures are cooler and after the sun has set. In either case the horses are obtaining the benefits of grazing even though they do not have access to grass for 24 hours a day.

A good rule of thumb is to figure an acre and a half of

Supplying your horse with a mineral block in his pasture and in his stall will assure he is able to take advantage of it at any time he wishes.

Avoid too many horses together in the same field. Two or three horses can provide valuable companionship to each other and avoid the problems of overgrazing a small area.

pasture to comfortably support each horse. If you have more horses per acre than this guideline, you may be in danger of overgrazing your pasture and the quality of the grass may suffer. Ideally you will also want to practice pasture rotation and allow fields to sit empty for some months of the year. Pasture rotation will also help you keep on top of the weed population and allow the area to rest from continual grazing. If your pastures are of high quality with few weeds and good quality grass, the figure of one and a half acres per horse may be too much. Knowing your situation and the status of your pastures will help you determine how many horses are suitable for your pasture. The condition of your horses (a hard keeper who can use the additional grass or a very well-covered horse who could stand to lose a few pounds) will also help you decide how many horses can safely inhabit your pasture. On our farm we have one pasture that we keep grazed down more than the others so it can be used as a diet pasture for those overly round individuals who don't need

hours on lush grass but enjoy nibbling.

There are a few points to keep in mind when putting your horses on pasture for part of the time. Lush, rich, tall grass is a horse's paradise, but if he is not used to being on grass you will need to slowly make this dietary change. If you turn out your horse on a lush pasture and allow him to eat to his heart's content, he will be at risk of colicking or foundering from the overabundance of grass. Slowly make these (and any) changes to his diet. If your horse has not been on pasture and you want to introduce him onto one, take your time. Allow him approximately one hour of grazing the first day or two, then return him to his paddock or stall. Increase the time on grass in small amounts, perhaps an hour or so a day, until you have gotten him used to this new situation, and soon you will be able to leave him on the pasture for as long as you wish. Taking a little time to ensure that the transition period is slow and cautious is always the best plan.

Some horses and ponies are very easy keepers, meaning

that they stay round and fat even on the smallest rations of feed. These equines are often the ones that can become obese when on a pasture of quality grass so you will want to carefully monitor their body condition. If you think that your horse is becoming too fat on grass, you may want to limit its time at pasture and supplement with some grass hay. Talk with your veterinarian about your horse's condition and try to settle upon the right balance of grass, hay, and/or concentrates that will keep your horse in optimum condition.

If you have broodmares with foals, then keeping them on pasture is one of the best things you can do for them. It takes a lot of food to keep a lactating mare from becoming too thin and good grass is a wonderful boost to her daily diet. In addition to the nutritional benefits for the broodmares, being turned out on a large pasture is fabulous for the foals because they have room to romp and play, a clean pasture to sleep in, and a picture-perfect environment in which to grow up.

SUPPLEMENTS

As we have already discussed, supplements of vitamins and minerals (with the exception of salt) are often unnecessary in your horse's diet, as so many of the needed nutrients are already provided through your horse's consumption of grass, hay, and concentrates. However, many horse owners are choosing to supplement. How do you know when and if you should consider this for your own horse?

There are many supplements on the market nowadays. There are supplements for improving coat condition, building bone, calming nervous horses, increasing fertility, improving hoof condition, and easing arthritis. There are also supplements for muscle soreness, improving digestion, and improving joint condition. Additionally, some horse owners feed corn oil to add fat to the horse's diet without adding the bulk of enormous amounts of grain. These are a sample of the types of supplements that are available to purchase and many horse owners feed multiple combinations of these products to their horses every day.

As a conscientious horse owner, you probably take great effort to provide the best you can for your horse, from his health care to his feeding program to his stabling. You may find a

Poisonous Plants

Before you turn your horses out on your pasture for the first time, take a stroll around your fields and keep an eye out for any poisonous plants that should be removed prior to horses inhabiting the area. Even after your horses are out grazing it's wise to continue to make periodic checks to make sure that unwanted plants and weeds are not encroaching your pasture. Don't overlook poisonous trees, even if they aren't directly in your pasture, as they may be a problem even from a distance. If the trees are nearby and the wind blows branches, leaves, and acorns into your pasture, these can still cause a problem. Below is a list of various weeds, plants, and trees that are poisonous to horses. Bear in mind that this is by no means an exhaustive list of poisonous plants, and there are many others which are harmful to horses, so do your research on any plant that concerns you.

Bracken	**Ragwort**
Buttercup	**Red Maple**
Castor Beans	**Rhododendron**
Hemlock	**Saint John's Wort**
Horsetail/Marestail	**Sorgum**
Loco Weed	**White Snakeroot**
Milkweed	**Yellow Star Thistle**
Nightshade	**Yew**
Oak	

During the winter months, you will have to supply your horses with large rations of hay to make up for the grass they would normally eat. You may observe your horse digging through the snow to find grass underneath, but don't expect this practice to benefit his nutritional needs in any way. The tiny amount of brown grass he might dig up will not be nearly enough to satisfy his daily requirements.

comforting feeling in feeding supplements to your horses, with the idea that they may help to maintain your horse's health and well being, and certainly can't do any harm, right?

Probably not, but it's still a good idea to discuss the use of supplements with your vet, who will be able to recommend the best supplements and help you decide whether or not you should include them in your horse's daily diet. Some companies produce pre-packaged supplements in small amounts for daily feeding and save you the time and effort of measuring each supplement every day. If your horse is on a daily dewormer, this can also be included in these pre-packaged supplement packets.

CHOOSING HAY

The actual feeding of concentrates can be relatively simple.

Once you have determined the type of concentrates that best suit your horse, all that remains is to purchase the necessary bags and measure out the feed. It seems pretty straightforward but choosing hay is an ongoing task that you will do with regularity for as long as you own horses. There are many types of hay and a vast range of quality and varieties of bale sizes to choose from. You might purchase absolutely perfect hay from a farmer one year and when you go to check out his hay the next year it may be dustier or less green. You need to continually keep your eyes open for good sources of quality horse hay.

RECOGNIZING GOOD QUALITY HAY

Regardless of the variety, all good hay should display the same characteristics.

A good smell. A sweet, fresh smell when you open the bale is always a sign of hay that was baled correctly. A

The cribbing this horse displays may be caused by the tension and excess energy from too little time spent grazing. Even exercise and turnout in a dry lot may not be enough, as horses can relax most easily when eating grass.

moldy bale or one that is leaning in that direction will have a rather earthy, musty smell.

Low dust. Very dusty hay is a sign it was baled too wet or was rained on in the field after cutting. All hay will show some dust, but large amounts are unhealthy to your horse.

A lot of leaves and fewer stems. Hay that was cut when it was too mature will have thicker, coarser stalks and less leaves. Hay that is too mature has lost the majority of its nutritional value.

Greenish color. Fresh hay has a definite greenish tint. Avoid brown, yellowish hay, which can be a sign of damage or low nutritional content. Gray or black patches are mold and must be avoided.

Few weeds. The hay should be consistent and free from other plants that at best may take up space in the bale while being of little nutritional value, and at worst may be possibly toxic.

No contamination. The hay should be free from foreign objects, such as sticks and trash. Also be aware of small animal carcasses (mice, snakes, frogs), which can be accidentally baled up with the hay. These are dangerous and may cause botulism. You should never feed bales that contain dead critters.

VARIETIES OF HAY

The type of hay you feed your horse will depend on your geographical location. Unless you transport hay from a distance, you will have to feed what is available in your area. There are many different varieties of hay that grow throughout the United States, but generally hay can be divided into two groups: grasses and legumes.

Grass hays are the only option if you are feeding free choice hay, as any other hay is too rich to be fed in large quan-

Nursing mares will do best on abundant quality pasture. The nutritional needs of broodmares are extremely demanding and there are few solutions better or more beautiful than keeping her and her foal at the liberty of a large expanse of quality grass.

tities. Grass hay comes in several varieties including orchard grass, timothy, bermuda, and brome. While these grasses do not have quite as much nutritional value as legume hays, they are a great type of roughage to feed to most horses.

Alfalfa, clover, and bird's-foot trefoil are types of legume hay and are considerably more nutritious than grass hay. However, legume hay must be fed sparingly, as overfeeding of such rich hay can cause colic or laminitis in some horses. Legumes can be good for horses that are hard keepers or for lactating broodmares.

Many horse owners prefer to feed a mixture of hays, such as an orchard grass and alfalfa mix. This is a great option because you can feed larger amounts (due to the grass content), while making the mix more nutritious and slightly more palatable (due to the alfalfa content).

SQUARE OR ROUND BALES?

Hay is generally available baled in two formats: square and round. In square bales the hay is compressed and tied into uniform rectangles that can be easily stacked. The hay is baled into sections or flakes that come apart when you open it. These divisions can make it easy to judge how much to feed your horse. Also, you can open a square bale and examine it all the way through at once and immediately notice if it is moldy or contaminated in any way before you've fed any to a horse. Square bales are reasonably light and typically weigh only 50 to 70 pounds, which makes for easy handling and storing.

Round bales are fed to horses in pasture situations where the horse is outside all or most of the time. Round bales are made by rolling the hay into a wheel-shaped stack about 4 to 6 feet wide and 4 to 6 feet tall. They usually

As always, the overall body condition of your horse should be watched carefully, particularly when grazing in pastures with abundant yield. Restricting his grazing times may help, but it might also encourage additional pasture growth.

Checking your horse's hay for mold or weeds before feeding is a quick way to ensure that you are giving him the best roughage possible.

Just because a poisonous plant or tree is not directly inside the pasture, don't underestimate your horse's ability to reach our for whatever he wants. Horses often seem determined to cause harm to themselves and are very clever.

weigh anywhere from 500 pounds to over a ton, depending on the hay variety and baling machine used to make it. They are considerably more difficult to move and store than their square cousins and require equipment such as a tractor with a front end loader. Round bales are more prone to mold and contamination than square bales because they are generally left outside in the weather. It is also impossible to inspect the hay before the horses eat it. But in a 24/7 pasture situation, some people prefer round bales because they can cut down on feeding time and allow the horse a free choice diet. Over time, particularly if there is more than one horse feeding, you will lose a large amount of hay that the horses tear off onto the ground, walk all over, and refuse to eat. Also, the herd's pecking order will determine how much, if any, the low horses are allowed to eat from the round bale. Some of your horses could suffer

if forced to diet by the bullies in the herd. The herd's daily habits must be observed and action taken if this occurs.

PROPER STORAGE OF HAY

Storing hay is all about one objective—keeping it dry. Dry hay is clean and healthy hay. You should never feed a moldy bale or even a partially moldy bale because it can cause illness or kill your horse. Rather than worry about moldy hay, one should simply take a few simple precautions to avoid mold.

Square bales are ideally stored indoors in a well venti-lated building, and everyone seems to have a different idea on how to stack it. Really there is no right or wrong way for every situation. Many people like to stack their hay on wooden pallets to help with ventilation and give the bottom bales a few inches of breathing room between them

Before purchasing a large quantity of hay, make sure that you are obtaining it from a reputable source and that the hay meets a high standard of quality. Have some extra helpers around if you're unloading a large amount.

From left to right: first cutting alfalfa/grass mix, first cutting grass, second cutting alfalfa/grass. Generally, a second or third cutting crop is softer, richer, and more abundant in nutritional value than first cuttings, which tend to be more stemmy and a less intense green. What you choose will depend on the body condition of your horse.

and the floor. Another method is to place the first few rows on their sides, as this is reported to also help avoid mold. Pallets are a must if the building has a dirt floor—otherwise the bottom bales will surely mold.

If the roof of the shelter where your hay is stored leaks it will be necessary to cover the entire load with plastic sheeting. This small precaution can save a huge amount of money if any unforeseen leakage should occur in the future. However, even with a sound roof, plastic may still be a good idea because it can protect the hay from birds should any decide to make your barn their "home." Just a few birds can make an enormous mess in a short time, which is a very bad thing for horses and can render all of your surface bales unusable.

If you live in a very cold climate you may want to think about not stacking your load directly against the walls. Although doing this can make the load more stable, frost during the winter months will undoubtedly form along the inside walls of the barn and potentially soak the adjacent bales in the spring when it melts.

If there are any windows in the building where the bales receive direct sunlight, you may notice the edges of those bales turn yellow. This is just bleaching from the sun and won't harm the hay. Even though it looks brown on the

outside, it will stay green on the inside and has no effect on the nutritional value.

Beyond these ideas keep in mind one other—please use common sense when building your pile. Don't stack the bales to a dangerous height just because you can, and be very careful when removing bales from the pile. Stacked hay is very heavy and the last thing you need is to create an avalanche.

STORING UNCURED HAY

Ventilation is especially important for fresh, green hay that has not been fully cured. Even though it will have been thoroughly dried before baling, uncured hay can still release a tremendous amount of moisture, which must be allowed to escape, otherwise it may cause mold or mildew in the bales. This is probably the only time plastic covering is not a good option because it will trap excess moisture and condensate under the plastic. On fair weather days, keep as many doors and windows open as possible to allow the humidity to escape.

A byproduct of the curing process is the steady release of heat as the hay dries. You can sometimes feel this on hay that has just been baled. If you reach your hand deep inside the bale you can feel its warmth. Although rare, one thing

Square bales are small enough to be managed by hand, which makes them a more standard choice for most horsemen and much easier for daily use in the barn.

A full winter's load of hay is safely stacked up high in the barn with early morning sunlight streaming through the windows. Scenes like these will be a daily joy and part of the general allure of farm and horse life.

to be aware of is that uncured hay also has the potential to start a fire through internal combustion. If it is stacked tightly together in large quantities, this heat generates inside the pile, is unable to escape and creates a possible fire hazard. Make a conscious effort to avoid stacking the bales firmly against each other. Give them a little breathing room. Again, these problems are eliminated once the hay as been properly dried.

Despite your best efforts, you will end up discarding a few bales over time. It's just part of the numbers involved. If you put up a stack of 600 to 1,000 bales, a few will still be moldy or dusty. Use your best sense and don't dwell on it. The increased cost among the remaining bales for the one you threw out it insignificant. The idea is to protect the majority of the bales from damage.

EVALUATING BODY CONDITION

Once you have decided upon an appropriate feeding program for your horses, it's easy to settle into a pattern of normalcy and confidence in your well-planned equine diet. While this is fine, it's always important to keep a close eye on the body condition of your horses, and be aware that you may need to tailor your feeding program from time to time.

While most people can recognize a horse that is too thin, and will realize that it is not a desirable condition, it can be harder for people to realize that an overweight horse is just as undesirable. Neither extreme is healthy. When you start out, have a feeding regimen that suits your horse and don't assume that this feeding program will be suitable forever. Seasonal changes have a major effect on the body condition of horses; as do other variables such as age, amount of exercise, whether the horse is being used for breeding, etc. You will constantly be tailoring the amount of hay and grain that you're feeding and adding more if the horse's weight seems to be falling off and reducing the amounts if the horse is overweight and not being exercised much.

As a horse owner, you must learn to recognize the subtleties of determining when your horse crosses over from one range to another. How you can tell if your horse is normal?

Generally, a horse in good body condition has enough body fat so that the ribs cannot be seen, but not so fat that the ribs cannot be easily felt. The prominent bones (withers, hips) should be moderately covered, not bony and protruding. Horses that are too thin are angular and their bodies seem to fall away from their topline. Obese horses have noticeable fat deposits, especially along their hindquarters and neck, which becomes extremely thick and cresty.

Let's say you want to figure out your horse's body weight. Since you won't be able to set your horse on your bathroom scale, there must be another option available and the most common is height-weight tape.

Height-weight tapes can be purchased from tack or feed stores and you can use one side to measure your horse's height (though it is not as accurate as a regular measuring stick) and the other side to measure the horse's weight. When measuring for weight wrap the tape around the horse's girth area and see where the two sides of the tape meet. This is the estimated weight of your horse. This inexpensive tool is a very handy thing to have in your tack room and is something you will use often.

There's also a formula you can use to figure the weight of your horse. Using a regular measuring tape, measure around your horse's girth (in the same place as you would use the height-weight tape) and record this number in inches. Then measure from your horse's point of shoulder to his point of buttocks and record this number in inches. Then use these measurements as follows:

Girth x girth x length, divided by the number 330

For instance, assuming a girth measurement of 67 inches and a body length measurement of 65 inches, you could calculate your horse's weight as follows:

(67 x 67 x 65) ÷ 330 = 884 pounds

The top of the hindquarters may become apple shaped with a crease down the middle and fat on either side.

If your horse is overweight you should try to reduce his weight so that he is in a condition more suitable for optimum health. In addition to reducing his food intake, you can also try to increase his exercise so that he burns additional calories and prevents them from being converted to fat.

If a horse is dangerously thin (completely emaciated, not just a little bony around the hips), you must proceed very carefully. Obviously, you will want to increase his daily feed ration. However, this must be done very slowly. A thin horse can be damaged by a too-rapid increase of food, particularly protein. If his digestive system has become accustomed to poor nutrition and small quantities of food, you will overwhelm his system if you rapidly introduce large amounts of rich food.

If you have two horses on the same diet and one is overly fat and the other overly thin, this may be due to the fact that the fat horse is more dominant than the thin horse. The dominant horse will hoard all of the available food and intimidate the thin horse to the point that he is unable to get his share of the feed. To prevent this situation from occurring, you may want to try feeding them separately so the thin horse is able to take his time and get his share of the nourishment without being bullied by the dominant horse. The other benefit from this separation is that the fat horse will be prevented from overeating.

HOUSING YOUR HORSE

A boarding stable may be the ideal solution for you after considering the advantages and disadvantages of keeping your horse there rather than on your own property.

When you purchase your first horse, one of the first decisions you'll face is where to I keep it. Basically, you have the choice of preparing a place on your own property to keep your horse or you can find a stable to board your horse. Both of these options have their benefits and drawbacks, and only you will be able to decide which option is the right one for you and your horse.

BOARDING

FINDING THE RIGHT FACILITY

Depending on your location, there may be several different boarding facilities to choose from in your area. You'll want to compare them closely to ensure that you choose the facility that offers the best quality care. A major key to choosing a boarding facility is cleanliness. You want to see clean stalls, clean paddocks, clean horses, and clean water buckets and troughs. The barn aisles should also be clean. Everything should appear well cared for and tidy.

Newer is not necessarily better, but if the facility is old and hasn't been maintained properly, there could be problems with rundown fencing and stalls with wear and tear. Check to be sure that the buildings, fencing, and stabling appear to be well-kept and in good condition.

Ask about the barn policies. Are the horses fed at regularly scheduled feeding times? How often is hay offered? How often are the water buckets and troughs checked, cleaned, and refilled? How many times a day is grain offered? How often are the stalls cleaned? If your horse is on a turnout schedule, how often will he be turned out and with how many other horses? Will your horse be used in the barn's lesson program, and if so, is there a discount on your monthly board bill to reflect this? Can you visit your horse any time, or are there posted barn hours and scheduled riding times? Who is in charge of the daily care of the horses, and who will be watching out for the daily welfare of your horse?

Look around thoroughly and ask other boarders for their feelings regarding the facility and staff. You can also get recommendations from local veterinarians, farriers, and feed-store owners. The facility can be beautiful, modern, and up-to-date, but you should also have a good rapport with the barn manager because this is a person you're going to be working closely with and seeing often.

The horses at the facility can give you insight to the care and attention that is given. They should be healthy with shiny coats (obviously this does not apply in the winter when they have heavy winter fur), and in good

Purchasing 24-hour turnout can represent a significant savings on your board bill. However, this is not always the best option, particularly if you will not be personally observing your horse on a daily basis. If he is constantly at liberty it may be difficult for others to evaluate his day-to-day condition.

weight. Don't discount the facility if you see one horse who seems a little thin. It could be an older horse who has trouble maintaining body condition. It's the general appearance of the horses on the whole that you're looking for.

Check the condition of the turnout areas. Is there grass, are they dry lots, or are the horses standing in mud? Are there trees for shade and three-sided shelters for protection against the elements? All of these factors are important to take into consideration when choosing a boarding facility.

The proximity of the facility to your home is another consideration. The closer you are to the barn, the more likely that you'll be able to spend time there. If you must drive a long distance every time you want to see or ride your horse, odds are that you'll have a hard time finding an opportunity to do so. Having a barn that's right around the

corner makes it infinitely more convenient to check in and see your horse on a regular, even daily, basis.

THE BENEFITS OF BOARDING

If you have only one horse, boarding him at a stable can allow him to enjoy the companionship of other horses, rather than spend his days alone in your backyard. Horses are herd-creatures and most are happier around other horses.

Another added benefit of boarding is that you can have frequent interaction with other horse owners and enthusiasts. You'll probably find that many people in your life are not particularly interested in your horse or hearing your stories about him, even though you may find the tales completely fascinating. It's nice to have some horse friends that you see frequently who are always interested in hearing

Boarding your horse will allow you to enjoy the advantages that the stable has to offer, such as this spacious, well-groomed riding arena.

about your horse's latest escapades or the things you've learned in your riding lessons, and are always ready to discuss the cost of saddles or helmets with you.

Boarding facilities often host special events that you can participate in, such as trail rides, holiday events, or clinics. These can be fun ways for you to spend time with your horse and your friends.

AVAILABILITY OF TRAINERS AND LESSONS

Many boarding facilities offer training and lessons, which can be one of the greatest benefits of boarding your horse. You're able to take advantage of the opportunity to obtain training for your horse and for yourself from the trainer at the barn. The training can sometimes be included in a lump sum board bill. You can arrange for one or more lessons per week, and there's great convenience in having your horse kept at the same facility where your training is conducted so you don't have to haul him back and forth from home to the barn in order to take your lessons!

It's very possible that you will want your horse to receive some professional training, especially if he is young and green. It's also possible that you may want to give him a refresher course even if he's older and has had previous training. If you're planning to ride, you want

A quality indoor arena will prove invaluable to you over time. It can be an especially convenient place to work a young or green horse since there are less distractions competing for his attention.

your horse to be as safe as possible and additional training is always beneficial.

It's also possible that even if your horse is well-trained and reliable, you may be the one in need of some education. Riding lessons can improve your knowledge, ability, and position, thereby increasing your safety and fun.

Having a resident trainer is very helpful in the day-to-day management and care of your horse. You may have questions about your horse and the trainer can provide valuable advice and assistance in situations where you're unsure about how to proceed. Behavioral problems, which you as a novice might find difficult to solve, can be simple for an experienced trainer to work through. Your trainer can also provide help in laying out your horse's feeding program, deworming and vaccination schedule, training

Housing your horse at a stable can put you in contact with other equine enthusiasts who share your passion. Everyday activities, such as trail riding, can become far more enjoyable when joined by the company of others who participate in your common interest.

and exercise regimens, and also offer advice on which turnout schedule will be most beneficial to your horse. Ultimately, the final decisions remain in your hands as the owner, but you may find lots of good advice and suggestions from talking with your trainer.

USE OF FACILITY

Try to compare what it might cost you to build a facility on your own property with the amenities that the boarding facility can offer. You could pay many years of board bills before you would ever come close to the costs of building a facility with stalls, an indoor arena, an outdoor arena, washracks, a tack room, a bathroom, a lounge, fenced paddocks, and outdoor shelters. By boarding your horse, you're gaining the benefits of these amenities without having such major expenditures from building your own, not to mention the impracticality of having such a facility on your property for only one horse.

Having the use of an indoor arena can help to further the training of your horse and the progress of your own

Riding lessons are highly recommended, even for those who have ridden in the past but may lack the complete understanding or fitness required for safe and effective horsemanship. Often there are horses belonging to the stable that will be available for you to use.

riding skills. If you don't have to cancel your lessons due to rain, snow, wind, or heat, you'll have much more consistent progress than you would if you had to frequently delay training due to weather extremes. An indoor arena is a major benefit to many boarders.

CARE WHEN YOU'RE AWAY

If you're headed out of town for an extended weekend, who will take care of your horse? Well, if he's stabled at a boarding facility, the answer is simple. The same individuals who care for him every day will be watching out for him even while you're away. His routine can remain the same, with the same feeding times, and knowledgeable horse people who are familiar with his personality, his likes, and his dislikes will care for him. This is a major benefit of boarding. It allows you more freedom of coming and going because the daily care goes on just the same whether you are there or not. This is in contrast to home horse care where you are completely responsible for the care of your horses at all times. If you need to be away, you have to find replacement help. It can be hard to find responsible, knowledgeable people to care for your horses while you are away. If they are unfamiliar with horses, they may not recognize if something is wrong, such as colic, sudden lameness, or if one of your horses isn't drinking. Similarly, they may not understand the danger of certain situations and can unintentionally put themselves—or your horse—in harm's way. Therefore, leaving your horse in the care of a boarding facility can put your mind at ease.

For the working horse owner who must be away from their horse for many hours each day, your horse may be able to have more feedings per day if he is kept at a boarding facility where there are people around all of the time, in addition to extra attention.

EASY ACCESS TO VETERINARY CARE AND FARRIERS

There's no need to spend time on the phone scheduling vet and farrier appointments when you're boarding at a barn. The barn owner is the one scheduling the appointments, and your horse will be taken care of when the

You may also take lessons using your own animal. Either way, the in-house trainer will be happy to work with you to achieve whatever your riding goals may be.

professional comes for his regularly scheduled visits. You're also saving on the farm call fees that are often charged by veterinarians and farriers when they come directly to your home barn.

If you don't feel comfortable using the farrier or veterinarian that the boarding facility uses and have your own preferences regarding your choice of professionals, you will have to arrange your own appointments and pay the farm call fee. You may also need to be on hand to handle your horse while the vet or farrier works, as the schedule might not coincide with that of the barn manager. Some barns have policies regarding the use of their own vets and farriers so it's important to understand these rules before you sign any boarding contracts, especially if you have a specific vet or farrier that you feel strongly about using or not using.

Regular chores at the stable will continue whether you are there or not, which gives you some flexibility in your schedule and the chance to own a horse without the possible time constraints of keeping him at home.

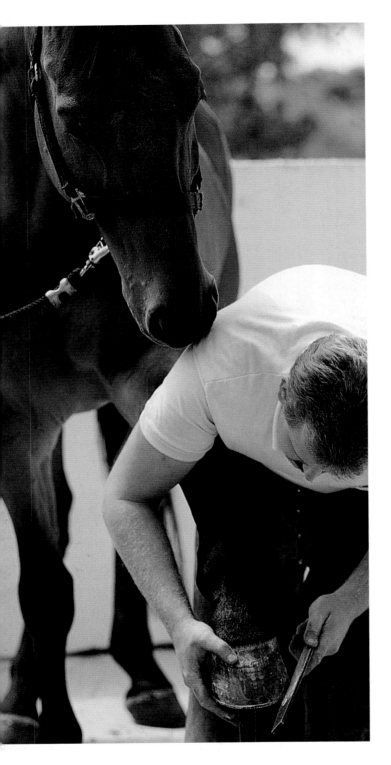

Convenient access to the services of horse-care professionals can be a major benefit to stabling at a boarding facility. You may also save money on farm call fees.

WHAT ABOUT COST?

We know what you're thinking. Boarding sounds great, the amenities sound fabulous, and I'd love not having to muck stalls, but what's the catch? Unfortunately, there is a catch, and that is the cost. At first glance, it may seem that boarding your horse may be an expensive choice, but you really can't make that determination until you have added up the costs of everything that you will be spending if you keep your horse at home. You'll probably find that although it is usually more cost-effective to keep your horse at home, you really must consider whether the benefits of boarding outweigh the additional costs. The cost of boarding will depend on a number of factors, as most barns offer a variety of different types of board.

Indoor (or stall) board is a complete package that includes everything: stabling at night and turnout during the day, stall cleaning, blanketing, multiple feedings per day, vitamins, handling for vet/farrier, etc. This is the most labor intensive for the barn staff and is usually the most expensive boarding option.

Pasture board is usually a less expensive option. It typically includes 24/7 turnout and an outdoor shelter but no stall. This option usually includes two feedings per day of hay, but typically doesn't include grain. Because the staff doesn't have to turn your horse out and bring him back in, and with no individualized feeding or stall cleaning, this option can be much cheaper than full board.

If your horse is well-trained, quiet, and gentle, there's a possibility that the barn manager may want to utilize your horse in the facility's lesson program with other student riders. This situation can be a benefit to you in more ways than one. You can sometimes get a reduced rate on your board bill if the barn is able to use your horse for lessons, which can add up significantly over time. Additionally, your horse will benefit from the extra exercise and riding time, especially if you're like most people and are very busy without excessive amounts of free time for riding yourself. Being used regularly for lessons helps keep your horse fit and in shape. Some questions to ask before allowing your horse to be used in a lesson program are what level or riders will be using my horse for lessons and who is liable if someone is injured while riding my horse.

Boarding with lessons can be a good choice for the competitive rider because you can pay for a certain amount of lessons per month in a bulk fee with your board bill. Occasionally you can obtain a slightly reduced rate if you book multiple lessons along with your board.

If you're looking for other ways to reduce your monthly board bill, you can inquire about the possibility of working off some of your board by helping around the barn. Perhaps you can clean stalls a couple of times a week or help with feeding during the evenings. Some barns are always looking for capable, responsible help and may be willing to adjust your board bill if you put in a number of hours around the farm.

HOME HORSE CARE

For many reasons you may decide that boarding your horse is not the best option for your lifestyle. Perhaps you feel that you can't justify the expense of the boarding facility, especially if you will not be utilizing the other benefits, such as the arenas, training, or riding lessons. Or perhaps you prefer being the person in control of your horse's daily care and value the ability to personally oversee his diet and health. Maybe there are no boarding facilities close enough to be convenient or maybe you enjoy looking out of your kitchen window and seeing your horse peacefully grazing.

Whatever your reasons, if you have the space and the desire to keep your horse at home, you will have joined the ranks of many horse owners who find great enjoyment in keeping their horses on their own property.

Before you make the decision to pursue home horse keeping, you'll want to be certain that your property is properly zoned for horses. Some areas have restrictions on the number of animals allowed per acre, while others prohibit all animals. If you live in a rural area there shouldn't be any such restrictions.

Ideally, if you're going to keep your horse at home, you will have some pasture available for grazing. Keeping horses on wooded acreage is not ideal, although it is done. Some owners will take down trees and make a clearing for their horses in the woods. While this is certainly workable, it's not as desirable as pasture with good grass for grazing.

Home horse keeping is a major commitment and you must understand that there are certain responsibilities that must be fulfilled each and every day. You'll be out in the barn to feed at least twice a day, if not three times, and you'll have a horse to groom and fences to maintain. The

A glorious and very satisfying sight: your own horse peacefully grazing in the fields adjacent to your home. For many horse owners, this is as good as it gets.

Obviously, chores will take up some time, but efficient work habits and organization can help you spend most of your time enjoying your horse.

stalls must be cleaned at least once a day, with all soiled bedding and manure removed and replaced with fresh bedding. Water buckets must be filled and refreshed, fly spray needs to be applied, and minor scrapes and injuries will need to be taken care of. Your horse needs daily turnout for exercise. You'll need to deworm your horse, arrange for vet and farrier appointments, and observe your horse's general state of health. You'll have to haul hay and make trips to the farm store for supplies, although these tasks are not altogether unenjoyable.

The enjoyment of constant companionship with your equine buddy will make up for these chores. You will be able to experience the daily delights of watching him graze, frolic, and then trot up for a scratch and pet when he sees you coming. It may be easier to find the time for riding, training, and generally enjoying his company when he is housed right outside your front door.

Keeping a horse on your own property puts you in full control of his pasture quality and turnout time. It gives you control you would not otherwise have when boarding elsewhere.

Although leaving the halter on may make a shy horse easier to catch, the benefits of doing so are not enough when compared to the possible problems. Here a young colt tries to bite the halter of his friend, which is a situation that might lead to entanglement or harm to each other.

PASTURE/TURNOUT

Having access to pasture is a real benefit to the majority of horses. In addition to it being an excellent source of roughage, most horses benefit physically and mentally from grazing. If your horse is obese or insulin resistant and shouldn't have access to grass, you'll want to limit his turnout to a dry lot instead of a grassy field. Obviously, your home acreage may not have the option of lush pastures to graze on and you may have to settle for turnout in a paddock without grass. It is still important for your horse to have turnout so that he may exercise and move around, whether he is able to graze or not.

The number of hours your horse spends in turnout will depend on your situation and circumstances. If you have a young horse that is still growing and full of energy, then the more hours he has at pasture the better. Whether or not you decide to stall him for part of the day or evening will depend on your schedule and personal preference. Easy keeper horses (ones that stay fat on air) usually benefit from having less time out at pasture, although turnout on a dry lot is perfectly acceptable. Sometimes your weather will dictate how much time you want your horse to spend outdoors. If the weather is bitterly cold and miserable, you might prefer to keep him in during the cold nights and early mornings. On the other hand, if it's mid-summer and hot, muggy, and buggy during the day, you might want to consider stalling your horse during the heat of the day and turning him out during the cool nights. If your amount of grass pasture is limited, you might want to limit the number of hours that your horse has access to it each day so that the pasture quality remains higher than it would if he was constantly grazing.

When leading a horse to the paddock, remember to navigate gates carefully. Allow yourself to pass through them ahead of the horse. Lead him through the paddock.

When you turn your horse out, it's best to remove the halter. It can be dangerous to have a horse loose in a pasture or paddock while wearing a halter, especially if the halter does not fit properly. The halter can get caught on something (a fence post or a trough, for example) or the horse might hook a foot through it, which is a very dangerous situation. Therefore, it's best not to turn out your horse while wearing a halter, but if you must do so, make sure that it is a properly fitting (not too large or small) halter made of leather, not nylon. Nylon will not snap in the event of a bad situation, but leather will. However, any halter, including a halter that

fits perfectly, can cause problems over time by chafing the hair off a horse's face, behind his ears, and across the bridge of his nose. An ill-fitting halter can even cause indentations where it presses too hard on a particular place. This is especially problematic with foals and young horses whose heads are still growing and changing at a rapid pace. Halters can quickly become too small and start to cause problems so it's wise not to leave them on for any longer than it takes to lead a horse to the paddock and back.

Most people who want to leave halters on their horses do so because they are fearful of not being able to catch the horse

It's not a good idea to turn your horse out while still wearing a halter. Avoid possible dangers and take the halter off before you turn him loose.

again. However, rather than avoiding the problem by leaving the halter on, it's better to try and remedy the situation by teaching your horse to be easily caught. It's understandable that you might be cautious about turning a new horse loose at pasture if you have no idea as to the ease of catching it again. In that case you might be more comfortable leaving on a properly fitting leather halter for a few hours until you have determined that it's possible to catch your horse again.

Teaching a horse to be easily caught requires some patience, a reward, and a routine. Offering a treat each time you catch your horse can quickly reinforce the idea that coming to you is a good idea and pleasant experience. The treat can be anything that your horse is fond of such as grain, hay, a handful of grass, a sugar cube, or a carrot. If you would rather not bring treats out to the pasture, you can have your horse's meal of hay or grain ready and waiting in his stall so that after catching him you immediately head to the barn for a snack. Horses quickly associate coming to you to be caught with the pleasant reward of a meal.

A good fan set in the hallway near an open door can go a long way toward maintaining good air quality and circulation throughout your stable.

Some horses will play hard to catch if you're getting ready to work them under saddle and arrive at the pasture to catch them at a "suspicious" time of day. If your horse knows that he comes in for grain every day at 6:00 p.m. and you're out to catch him at 3:00 p.m., he may decide not to bother going in to work that day. This is where a treat in your pocket can help override those feelings of "I'm not ready to go in yet."

STABLING

A very exciting part of horse ownership is the planning of your barn. Whether you are going to build a brand new barn on your property or merely renovate an existing structure, the possibilities are endless for tailoring your

This is a large, clean box stall with fresh shavings and chew guards installed. It is a delightful scene to make any horse feel at home. Steel bars in front of the windows are always a good idea.

Wood shavings can also be compressed into bags. Bulk delivery is generally a cheaper solution but requires a large area of specific storage. Bagged shavings are easier to store and maneuver, but you will have to pay extra for this convenience.

Wood shavings can be purchased in bulk loads, such as the one shown here, which arrived in a semi-load at a boarding stable.

barn to your specific needs. On our farm, our main barn is a former dairy barn that was built in 1932. We have since converted it to a horse barn with eight stalls, hay storage, and a heated area for our water source to stay warm in the winter. We have also converted a vintage stone garage into a two-stall barn that houses our stallions and has custom-made stall components to fit the building's unusual

dimensions. Many existing buildings can be renovated to house horses which can possibly save you money versus building a new facility. If you don't have an existing building to renovate, you may decide that a brand new barn is the right option for you. The size of your new barn will depend on your specific needs and the number of stalls you wish to have in it, as well as your budget for the project. A four-stall barn with a tack room and wash stall might be the perfect size for your farm, or you might opt for an eight-stall barn and forego the wash stall. You might want to look into the pre-engineered barns that are manufactured by companies such as Morton Buildings and MD Barns.

Cleaning stalls is a task best tackled on a daily basis. This may seem like a lot of extra work since the stall might not seem too dirty after one night. But the key to keeping the job easy is consistency. If the stall is always brought back to good condition at the same time every day, it will never become an enormous task and will be a healthier environment for your horse, too.

STALLS

Obviously, your horse doesn't need to have a stall to survive; many horses are kept quite happily and comfortably in pastures with run-in sheds and never have a stall. However, there are many reasons that you would want to provide a stall for your horse. Extreme weather conditions are a major consideration. If you live in a climate where the weather gets dangerously cold (or dangerously hot), a stall is a very beneficial accommodation for your horse so he can get away from the extreme temperatures. You might

Pelleted bedding packs a lot of potential absorption into a small package. However, they require a bit more human effort when first used to ensure a quality bed.

enjoy keeping your horse in a stall in order to have easy access to your horse when you decide to ride. You may decide to utilize it as a place for the farrier to work or the vet to examine your horse. Many horses come to view their stalls as their "safe haven," and will often wait at the paddock gate for you to put them back inside. Their stalls represent safety, peace, food, and water, all very important things to a horse!

STALL SIZE AND SAFETY

Your box stalls must be large enough for the horse to move around freely and have ample room to lay down without being too close to a wall. Horses who attempt to lay down in a stall that is too small for them are in danger of becoming cast, which is when they are so close to a wall they can't roll back over and they become trapped. Also, some horses who are kept in stalls that are too small may become nervous from the confined, uncomfortable feeling.

There are several manure management methods, some of which may benefit others who are interested in obtaining fertilizer for gardens or other agricultural uses.

Think of all the money you're saving on gym memberships! The light physical work of daily stable maintenance can benefit you in more ways than one. In our barn, we have a sign that says, "Around here, the muscles are sore from work, not the gym!" Take pride in your hard work.

Even horses with access to naturally occurring drinking sources will need additional water options. It's always a good idea to provide water buckets or troughs in your pastures and stalls.

That said, unless your horse is a very large breed, a 12 x 12-foot stall is probably large enough to accommodate your horse, with a 10 x 10-foot stall being acceptable for a smaller animal or pony. (A foaling stall will need to be much larger, 14 x 14 feet or larger, as is discussed in Chapter 7.)

Ideally, the floor beneath the stall should be dirt, although concrete is a possible solution as long as you provide good bedding. Any stall flooring should be covered with rubber matting—tightly fitted to prevent leakage or heaving—and then covered with your choice of bedding.

Water buckets should be hung at a comfortable drinking height for the horse without being so low that he could catch a foot in the bucket. Hay racks, if used, should be hung slightly below the horse's eye level to avoid him having to reach up to eat, which can cause him to develop undesirable muscle build-up along the front of his neck.

If the stall is primarily built of wood you will want to install chew guards along the straight edges where your horse will be able to get his front teeth on. Some horses love to chew wood and can get into the habit of it, which if left unchecked can lead to cribbing. You will want to discourage the chewing right away. Just be sure that any chew guards put into place are safe and securely attached without any sharp edges.

Always remove your horse's halter when he is in his stall. You will be able to catch him in a stall and leaving it on is simply lazy at best and possibly dangerous at worst.

Garden hoses that are specifically rated for drinking use can be employed about your stable to easily deliver water to your horses from long distances.

Horses rub themselves on stall walls and lay down and roll. There are many things that a halter can snag on and leave a horse trapped for hours before anyone realizes that there is a problem.

All stalls must be checked for safety hazards, such as sharp edges or protruding nails, and the appropriate repairs should be made when necessary. Windows should always be covered with bars or a grid to prevent accidental breakage. If possible, have them placed as high as possible, while remaining convenient, to help prevent accidental breakage.

Automatic waterers, whether installed indoors or out, can be a time- and labor-saving system. Just be sure to introduce the horses to it beforehand to ensure they are used to the noise and motion and will not be afraid to drink from it.

If the electrical current is not particularly strong, it may become necessary to remove grass from underneath the lowest wire to keep it from shorting out. Remember to remove any grass clippings from the animals reach when doing so.

Wood fencing is a charming, traditional addition to any equestrian facility. Often the sight of such a fence brings positive feelings and memories to people who are not familiar with horses. Note the electric wire running along the top to encourage fence respect.

STABLE VENTILATION

Good air movement throughout your stable is extremely important. A cool but dry stable is much healthier for horses than a warm one that is closed up and humid. Good ventilation controls odor, condensation, mold, mildew, and dust, all of which can create problems.

Ammonia fumes from stalls and bedding can build up and irritate the eyes and respiratory system of a horse, as can dust from bedding, hay, dirt, and life in the barn. Moisture build-up from the horses' own breathing must be exchanged with fresh air. A damp stable will allow bacteria to grow, which can also lead to respiratory problems.

Good ventilation can be achieved by arranging doors and windows at cross points in the building, which allows fresh air to pass easily through the middle of the barn while allowing humidity to escape. Stall walls and doors that are

Electrical tape fencing is easy to install and highly visible to horses, who can sometimes have trouble determining the exact location of a thinner wire.

built with grids or bars instead of solid partitions can also help bring fresh air to your horse and remove stale odors. High ceilings and exhaust fans placed at the ends of hallways or on the peaks of roofs are other methods used to increase ventilation, as well as a good box fan safely placed outside your horse's stall to make him a bit more comfortable on a very hot, humid day.

BEDDING OPTIONS
Wood shavings are a good choice for bedding and widely available either loose in a bulk load or bagged. The best shavings have been well filtered to eliminate dust. If you cut open a bag and the entire stall fills with dust it will irritate your horse's eyes and airways whenever he moves about the stall and stirs up more dust. Good quality shavings are clean, consistent, and made of pine. Never use walnut shavings because they have been blamed for laminitis in horses.

High quality vinyl fences will not rot and should last a long time if not abused. But horses can be rough on fencing and a sturdier, more substantial option may be better.

Safety Around the Barn

Safety should always be taken seriously around the barn, with special consideration for fire safety. It's very important to have basic farm rules about the storage of fuels and the dangers of smoking in an equine environment. This can help protect your buildings and horses from possible trouble.

A fire extinguisher kept in an easily accessible locations is a great plan. Inform others of the chosen positions of your fire extinguishers so that everyone knows where they are.

For the safety of any visitors to your farm, you should always mark electric fences in highly visible places so that everyone is aware that the fence is under charge.

Corral panels are great for round pens and make safe places for the early training of young horses. Even the name corral brings back wonderful images of cowboys and Western horsemanship. You might also use corral panels as a sturdy but inexpensive private paddock for a stallion or stud colt, as seen here.

Wood shavings are very absorbent, easy to clean, and fairly cost-effective.

Saw dust is not recommended because it creates a lot of dust, which can cause the same problems discussed above. Even though it can be extremely absorbent, it is a poor trade off for the potential hazards that arise from too much dust.

Straw has traditionally been used as horse bedding and is still a dependable choice. While not as absorbent as other methods, a good straw bed is soft, clean, and when baled properly, it is fairly dust-free. Straw that is baled too wet or was rained on may be very dusty and possibly moldy and should be avoided. Straw is a bit more difficult to clean than shavings because it is very bulky. You will find yourself removing large quantities of soiled bedding, even though it is very light and

easy to move! Oat straw is probably the best choice for horses, although wheat and barley are also commonly used. Straw of any variety is typically inexpensive and gives you a large amount of bedding per bale for a price cheaper than shavings. However, straw bedding becomes damp very quickly and will take you considerably longer to clean, thus increasing your time spent on stall cleaning and manure removal.

Another product that has come into use for horses is pelleted bedding. This product is made up of compressed wood particles in pellet form. The bags take up smaller space in your storage building. The pellets expand upon contact with water and continue to expand and absorb as they are used. It is a more expensive type of bedding and it can be difficult to create a thick, comfortable bed for your horse.

Regular observation and maintenance of whatever fence you choose will go a long way in preventing potential problems before they become a threat or nuisance.

One other option available is paper bedding, but this is not a favored choice among horse owners. Paper bedding is messier, less absorbent, and not nearly as attractive as a stall filled with pine shavings. Bedding from paper is more prone to the growth of fungus and mushrooms, which is certainly not a desirable side effect of your horse's bedding!

MUCKING/CLEANING STALLS

Cleaning stalls is really not as bad as many people would lead you to believe. It's really not as time-consuming or disgusting as some people think. You are not going to spend every waking hour sifting through dirty bedding. You can probably figure on devoting 5 to 10 minutes to each stall twice a day. Once you've developed a system, you'll be able to breeze through your barn quite efficiently.

Barbed wire is a fence that is best taken down, not put up! Even if you do not pasture your horse behind it, any sections of this type of fence should be removed to protect people and riders of horses who may trail ride on your property.

Sometimes people will spend a large amount of time and money installing the perfect, secure, strong fence, and then close it off with a pitiful gate tied shut with baling twine. Go the extra mile and find a decent gate that can be secured with a large snap or other device the horse will not be able to easily break or open.

Ideally your horses will be turned out to paddock or pasture so that the stalls are empty while you clean them. Trying to clean a stall around your horse is inefficient and not safe. If you are unable to turn out your horse while you clean the stall you might want to consider having someone else halter and hold him out of the way, or perhaps you can put him in crossties while you work, but your horse will have to be trained to stand in the crossties before you attempt this.

The equipment you'll need to clean a stall is a stall pick/pitchfork and a wheelbarrow or muck bucket. Start by removing any visible manure piles and proceed around the stall. Sift through the bedding to look for any hidden piles or loose droppings that you might have missed. You'll also be on the lookout for wet bedding that needs to be removed. Once that is completed, re-spread the remaining dry bedding throughout the stall and add additional new

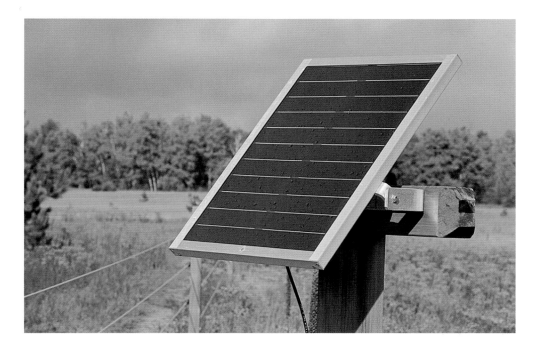

Depending on the size of the field, you may be able to obtain a successful electrical current using solar power. Be aware of the changing seasons and adjust the angle of the panel as needed to absorb the maximum charge. If you cannot keep a decent charge, an electric fence controller powered by a weather-proof power outlet can provide a reliable charge no matter what the weather or time of year.

These two young mares were just introduced into the same pasture for the first time. After a few moments of mutual nose-sniffing, things may look fine.

bedding if necessary. You'll want to completely empty the stall of all the bedding and refresh it at least once a week.

If your horses are turned out at pasture and aren't spending much time in their stalls, the cleaning required will be minimal and you'll pick out a pile here and there. But if they are spending several hours a day (or all night) in the stall, you'll need to clean it thoroughly each day.

This chore doesn't have to be drudgery either. Keep a radio or CD player out in the barn and listen to some music while you work or use the time to do some uninterrupted thinking. Out in the barn you are free from the distractions of the telephone and television and you can enjoy some quiet time. You might just enjoy it enough that you forget you're shoveling manure.

The evening walk back from the barn on a frigid wintery evening can be a very peaceful journey, especially when the sky is clear and the stars are abundant or the Northern Lights are glowing in a spectacular blur of color. Times like these make your stable chores seem so much more worthwhile and will increase the happiness in your chosen lifestyle.

MANURE MANAGEMENT

Sooner or later as you clean stalls and dispose of the soiled bedding and manure you're going to have your very own manure pile. With one horse, it may take some time before the pile is large enough to need attention, but if you have multiple horses, the pile can accumulate to an impressive

The apparent affections for one another may quickly deteriorate if one mare decides to seek dominance over the other and begins to cause trouble.

size in a short amount of time. The average horse produces eight piles of manure a day, so multiplying that figure by your number of horses, times 30 days, will give you an idea of the amount of manure you will deal with per month. And that's not counting the wet and soiled bedding that will also be on your manure pile!

How you dispose of the manure pile will depend on your location. Horse owners in suburban locations may have zoning restrictions with regard to where the manure pile can be located on their property. Rural horse owners may wish to spread the manure on a field as fertilizer, but fresh manure cannot be used. The manure must decompose for several weeks prior to being used on pasture.

Many landscape companies will remove your manure pile with their equipment. Sometimes you can make an arrangement to trade their labor in removing the manure in exchange for them using the manure at no cost.

Advertising in the newspaper can sometimes put you in contact with gardeners who are eager to haul away some of your manure for their gardens. Another option, particularly if you only have a horse or two, is to dump the manure and bedding into a trash receptacle and pay the garbage company to haul it away weekly or monthly. Ask your garbage company beforehand to be certain that it's acceptable for you to use the receptacle in this fashion.

You may have to pay someone to come in periodically and dispose of the pile. Horse owners on small acreage

If your horses spend a great deal of time using an outdoor shelter, you will need to regularly maintain its cleanliness, just the same as an indoor stall.

parcels are particularly limited in the amount of space available for their manure pile, and paying for its removal may be an expense that you frequently incur.

WATER

Your horse will need a water source wherever it is spending time, whether it's in a stall or out in a paddock. In the stall, your system may be as simple as a water bucket hung on the wall that is regularly washed and refilled with fresh water. You can also use an automatic watering system, which is very convenient as it supplies fresh water continuously. This type of system has a drawback because it's impossible to discern how much water the horse is consuming if the system is not equipped with a water consumption meter.

Out in the field you can use an automatic watering system or have large water troughs. These have the benefit of providing large quantities of water and don't need regular refilling throughout the day. However, they are more cumbersome to clean and you may have difficulty with bird droppings. You can also use regular water buckets, which are easy to carry and clean. These don't have the capacity that a water trough does so you will have to either supply several buckets or be prepared to refill at intervals throughout the day.

Natural water sources, such as ponds, are picturesque and serene in appearance but they are not completely reliable. Many horses prefer water from troughs or buckets because the pond water may not be to their liking due to

Horses who must endure a severe winter will need some sort of shelter to help them get out of the snow and wind. Doing so will keep them happy and healthy. Watching horses play in the snow is a beautiful thing.

possible contamination from other creatures who may also frequent the water source.

FENCING

Your horse is comfortably settled in his nice box stall filled with fresh shavings and has access to plenty of water and good quality hay. Now it's time to turn your attention to fencing. When it's time to turn him out, you'll need to put up a sturdy fence. There are several fencing options available with varying amounts of maintenance and costs to fit a variety of budgets.

WOOD BOARD

Board fencing is the traditional form of fencing for horses. It's safe, attractive, and durable when maintained properly. It's also expensive and requires regular attention (for painting and replacing boards, for example). Horses are notorious for chewing on expensive board fences. Many horse owners like to enhance the safety of their board fences by running a single hot wire of electric that is attached to the board around the perimeter of the fence. This increases your horse's respect for the fence, while maintaining the classic look of board fencing.

ELECTRIC

While not as aesthetically pleasing as other types of fencing, electric fencing is a popular option for many horse owners. Electric fencing is generally a very safe type of fencing that horses respect. This is a less expensive option, as there can be further distance in between posts and there is far less maintenance involved than with wood fencing. Obviously horses will not chew on electric fencing, although they may chew on the posts if they are wooden.

Your horse will need to consume more roughage during the winter months when his body is in need of additional calories to stay warm. The very motion and digestion of eating will help keep his body comfortable.

A sure sign that winter is on its way is frozen water in a bucket left hanging overnight. You will need to make sure your horse continues to have a reliable water source even in harsh conditions.

Electric fencing comes in a few varieties, including wire, braided rope, and tape. The safest of these are the rope and tape, as wire is less forgiving in the event of entanglement. There are many factors that can influence the effectiveness of your electric fence—proper grounding, correct choice of low impedance charger, and the size of the pasture being electrified. You will also need to regularly remove weeds or tall grass that can grow up alongside the fence and short out your electrical current. In addition, you will need to decide if a solar charger will be suitable for your needs or if you will install a higher rated system. Generally solar chargers work well with small fields that are not near any convenient power source. However, these can be fairly unreliable, especially in the winter when the sun is less intense or during long periods of cloudy weather.

VINYL

Vinyl fencing has several advantages. It's less expensive than wood, horses won't chew on it, and there is no maintenance. However, it can be rather flimsy and won't stand up to aggressive use if horses try to reach through or rub on it, so it may not be an effective solution for a pasture situation. These fences sometimes work better for riding arenas where you would like a more decorative fence without the cost of wood. As with wood, vinyl fence can be reinforced with a strand of electric wire to discourage horses from tampering with the fence.

CORRAL PANELS

These panels are portable, easy to move around, and allow you to set up small paddocks or round pens wherever you desire. Corral panel fences are not as expensive as some of the other types of fences. They can lack durability as horses can (over time) push and bend the panels, rendering them unusable. Depending on the type of finish, the paint can chip off and the panels may rust, but good quality panels retain their finish and do not require painting or additional maintenance.

BARBED WIRE

Barbed wire fencing might have been fine on Uncle Albert's farm when you were a child, but it is definitely discouraged for use as horse fencing nowadays. If there's barbed wire fence already installed on your property, you should consider removing it and replacing it with a safer form of horse fencing.

INTRODUCING A HORSE TO ITS COMPANION(S)

Introducing two horses can be a nail-biting endeavor for some owners. Making the transition smoothly takes a watchful eye, an understanding of herd dynamics and hierarchy, and a bucket of grain in case you decide that you need to get somebody out of there fast.

Let's assume that you have purchased two horses, a mare and a gelding, from the same farm. They are old friends, have been turned out together for several years, and you have had no

Fallen snow collecting on his back and not melting is a sign that your horse is successfully staying warm, as it shows his fur coat is doing its job and retaining body heat inside and not allowing it to escape and melt the snow.

troubles with them being turned out together in the pasture at your farm. You know very well that your mare is the one in charge. She tells your gelding where to go, when to go there, and how long to stay, and he follows her orders quite happily and without question. As long as you distribute their hay piles separately, they are both happy and they comfortably exist together.

But let's say that you've just made an additional purchase, a 12-year-old broodmare. This mare comes from a farm where she was the boss mare in a field of eight other mares. While she is quiet and docile for you to handle, you're not sure what she's going to do once she's turned out with your other horses. Your other mare and gelding are happily grazing in their pasture and you're ready to add the new mare into the mix. What's going to happen?

If you're lucky, maybe nothing. If your horses are gentle souls they may choose to ignore one another and spend the day grazing at opposite ends of the pasture, pretending that they don't see each other. This does happen, especially if your horses have laid back personalities.

However, as we are dealing with boss mares, they probably won't be able to ignore each other. There could be some chasing involved as they try to establish a sense of hierarchy. There may even be kicking and biting as your mares work out their differences. It should be quickly clear who is going to be the dominant mare. In all likelihood, the gelding will try to melt into the woodwork. It would be unlikely that he would try to obtain the head position of the herd, especially as he is already under the command of your original mare.

We had an elderly broodmare who would literally scream at the top of her lungs if any other new mare came near her. Every other horse on the property thought she was insane and kept a wide berth around her. This went on for several years until we purchased another new mare and added her into the group. For whatever reason, the old mare immediately resigned her position and would go anywhere or do anything that the new mare ordered her to do. This was without any scuffling or chasing whatsoever and was one of the easiest introductions ever accomplished.

In the event that the horses you've introduced just don't get along (i.e., they have chased, bit, and kicked each other for an hour without any sign of stopping), it might be wise to separate them for awhile and try again another day. Catch somebody and let things calm down for a while. There's not much to be gained from continual fighting, and things might be better the next time you try.

Another scenario that you might encounter is when all appears to be well at first, but then the horses begin irritating each other days or weeks later. Some horses aren't able to learn to get along and they may periodically torment each other (occasionally tension will increase when the mares are in heat). If you find that a pair of horses aren't working out together, it's best to do some rearranging until you find a better situation. Similarly, if you have a pair or group of horses that really interact well together and live peacefully, don't change anything! If they are happy, you are happy.

OUTDOOR SHELTERS

Whether your horses are turned out 24/7 or whether they spend some time in stalls, you're going to want some sort of an outdoor shelter in your pasture or paddock area. A three-sided shelter provides a good place for your horses to escape from weather extremes. It offers protection from rain and snow, a windbreak during violent winds, a shady place to escape the hot summer sun, and a place to get away from flies and other annoying bugs.

Your shelter should be large enough to amply house all of the horses in that particular pasture. You want there to be plenty of space for all of them to comfortably rest inside the shelter without having the dominant horses keep the meeker ones from entering.

The shelter will need to be cleaned out regularly with all of the manure removed so that the horses are standing in clean, dry footing and not in a mixture of mud and muck.

WINTER CARE OF HORSES

Harsh winter weather can add a whole new dimension to home horse keeping! The schedule and routine that works perfectly well in the warmer months may be totally unworkable in the winter time. Winter brings added concerns to the horse owner and you'll have to be vigilant in order to be sure that your horses come through winter in good condition.

It's natural, in your kind concern for your horse, to worry about him being too cold in the winter temperatures. In reality most horses are well provided for with their thick winter fur and additional blanketing is usually unnecessary. In most situations their natural winter fur is more than ample. However, there are unique situations where you might want to consider blanketing, such as an aged horse with difficulties in maintaining body temperature, or a horse that was recently transported from a warmer climate. For these circumstances there are a variety of blanketing options ranging from light sheets to keep your horse dry to heavy fleece-lined blankets designed for extra warmth. If you decide to blanket your horse, you'll want to remove the blanket regularly to check for any abrasions or injuries that you would otherwise miss underneath the blanket. You'll also want to regularly evaluate your horse's weight and condition, as you might miss the fact that he's slowly losing weight since you're not seeing his body every day while blanketed.

Winter fur also can be deceiving in regards to your horse's weight and condition. Fur creates an illusion of fat and you can't really get an accurate idea of your horse's body condition just by looking at him. Feel along his back and hips and along the shoulder area. Check to see if you can feel the ribs easily or if they are nicely covered by a layer of fat. Don't let the thick winter fur deceive you into thinking that your horse is fatter than he really is.

It's important to understand that even with his winter fur, your horse will be burning far more calories in the winter than he does in the summer. This is because the body must work hard at staying warm during the frigid temperatures. You will probably need to increase his daily rations of hay and grain during the winter months. Providing plenty of hay is particularly important.

If your horse has not developed a thick coat and seems to be suffering from extreme winter temperatures, you may wish to invest in a rug to help him retain his body heat.

This is why you may need to increase his daily rations of both hay and grain during the winter months. This is particularly true of hay. You may be surprised at the amount of food necessary to keep your horse in good weight during the winter. If your climate receives large amounts of snow, grazing won't be possibile for several months so his entire consumption of roughage must come from the hay you provide.

Even though shelter is very important at any time of the year, it's absolutely vital during the winter months. Your horse must have a place to get out of the snow, wind, and harsh conditions. This can be a three-sided loafing shed or a box stall in your barn.

Water consumption can go down drastically in the winter, as most horses do not care for the taste of very cold water and obviously cannot drink at all if the water is frozen. Decreased water consumption can lead to increased risk of impaction colic so you'll want to carefully monitor how much your horse is drinking. Be sure that you keep fresh (unfrozen) water in front of him as much of the time as possible. It can be difficult to do this when the temperatures are far below freezing and the water freezes so quickly, but it really is very important to do your best. Heated water buckets can be a very helpful accessory and will save you hours of removing ice from the water buckets, in addition to keeping the water warmer and more palatable for your horse. Make sure the bucket is hung in such a way that the cord is completely inaccessible to the horse.

It's a good idea to keep a salt block in your horse's stall or in his turnout paddock. Licking the block will encourage him to drink more water. Some people may tell you that water doesn't need to be provided in the winter because the horse's water intake can be supplied by eating snow. This is not good practice as the amount of water ingested through eating snow is so minimal it would be very difficult for a horse to remain hydrated if no other water source is provided.

If you are riding your horse during the winter (or if he is playing hard during his turnout time) and he becomes hot and sweaty, you should never return him to his stall in that condition. Even though he is hot at present, the cold temperatures will soon begin to freeze the dampness of his coat and possibly cause a chill. Be sure to walk him long enough to cool him out completely and then towel off any remaining dampness in his coat before returning him to his stall. It is a good idea to put a polar fleece cooler on your horse while you are walking him in the winter.

Snow and ice can pack into and freeze in your horse's hooves and make them very uncomfortable when walking. It also makes a horse more prone to slipping and falling, especially when walking on the barn floor or across a driveway. Check your horse's hooves daily and remove the frozen ice balls that build up in his hooves. Walk slowly when leading your horse. You don't want to slip or have your horse slip and fall.

Horses are at increased risk of some health conditions during the winter, these include thrush and rain rot, both caused by wet conditions. Also, you'll want to be certain that in your dedication to your horse that you don't forget to take care of yourself. Dress warmly in layers of clothing and warm boots and keep exposed skin to a minimum. Do your work quickly and efficiently. Don't be outside any longer than you have to. If you need to take a break and warm up inside for a few minutes, do so. Don't keep pushing yourself when the conditions are perilous. Take care of your essential chores and don't do anything unnecessary.

THE HEALTHY HORSE

Flies of various types can irritate your horse to the point that he cannot even enjoy his time outside and may turn the simple enjoyment of grazing into a miserable experience for him. There are several methods available for discouraging these pests.

Taking good care of your horse involves so many different aspects, from your feeding program and your stabling situation to training, handling, and general health care. Every horse owner wants his or her horse to be healthy and well cared for, and there are some basic things that you can do to help your horse to stay healthy.

BASIC HEALTH CARE

The best thing you can do with regard to your horse's health care is to choose a knowledgeable veterinarian, preferably one who specializes in equines. A vet's advice and suggestions will be invaluable as you learn more about your horse's health-care needs. Don't be afraid to ask questions. No question is a silly question when you are trying to make the best decisions for your horse's health. As you read about and research these topics on your own (vaccinations, deworming, exercise, etc.), you may feel a little overwhelmed by the amount of information and differing ideas that you encounter. Quite possibly you will get differing opinions from your horse friends on topics such as hoof trimming. Some may trim their horses' hooves every 6 weeks and others may never trim them at all. Some will vaccinate their horses for West Nile virus twice a year, others will vaccinate once a year, and others never do. Who

is right and how do you know what's right for your horse? Your veterinarian can help you to establish a health-care routine that is proper for your horse and you can feel confident in knowing that you're doing your best to make the right choices for your horse.

IMMUNIZATIONS

You will definitely want to consult your veterinarian on the proper immunization schedule for your horse, as the specific vaccinations given will vary depending upon your horse's personal needs and your region. For instance, horses in some areas are vaccinated for Potomac horse fever, while horses in other areas are not but might be vaccinated for other diseases endemic to their area.

Generally, it is recommended that all horses receive the following immunizations annually: tetanus toxoid, Eastern equine encephalomyelitis (EEE), Western equine encephalomyelitis (WEE), equine influenza, and rhinopneumonitis (EHV-1 and EHV-4). Additional boosters for equine influenza may be necessary if your horse is being shown and/or exposed to other horses frequently. On the

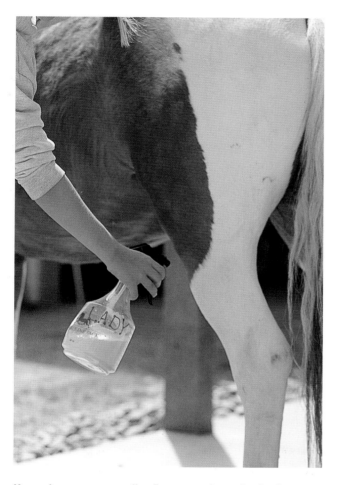

Here a horse owner applies fly spray to her animal using a simple spray bottle of the variety sold in many pet supply stores. Applying some before your ride on a hot day will probably make it more enjoyable.

end of mosquito season. If you live in an area where mosquitoes are prevalent you'll want to be sure that you keep your horse's vaccinations current, as your horse may not be under optimal protection after a few months.

For pregnant mares, there is a specific vaccination schedule that you will want to discuss with your veterinarian. See more on this subject in Chapter 7.

INSECT PARASITES

There are a number of insect parasites that are troublesome to horses (and humans!) including flies, ticks, lice, mites, and mosquitoes. None of these creatures are embraced by the horse owner and you will likely spend a good amount of time trying to eradicate these pests from your horse's environment.

FLIES

There are many varieties of flies including the stable fly, bot fly, horse fly, deer fly, horn fly, and the dreadful black fly. Flies are particularly annoying because they are so prevalent. While many are relatively harmless, especially those that do not bite, they are still a problem because of their annoyance to horses and their ability to transmit diseases, including equine infectious anemia. Additionally, some horses are allergic to flies.

Control of flies can be accomplished through the use of insecticides and repellents. The safest insecticides are those that include the ingredient pyrethrin. They are the most gentle on horses and are safer for frequent use. Repellents only repel the insects and do not kill them, but repellents do offer good protection from flies and mosquitoes.

There are a number of ways to try and decrease the general fly population around your barn or farm. One of the most important is the prompt removal of manure from stalls or paddocks. It's a good idea to keep your manure pile away from where your horses are housed. You can also use chemicals and sprays to help control your fly population, but it's important to make sure that any chemicals you use are safe for use around horses and thoroughly follow all application instructions.

Air movement is helpful in decreasing the number of flies in your barn, so keep the windows open (it's especially

other hand, if your horse is kept at home, isn't being shown, and has no contact with other horses, then annual vaccinations for equine influenza and rhinopneumonitis may not be necessary. Additional vaccines that you may want to discuss with your veterinarian include: rabies, strangles, Venezuelan equine encephalomyelitis (VEE), Potomac horse fever, equine viral arteritis, botulism, rhinopneumonitis (EHV-1 and EHV-4), and EPM. Horses traveling to endemic areas, such as the southern United States, might need boosters for EEE, WEE, and WNV.

In the fall, you may need to give boosters for vaccines that protect against mosquito-borne diseases because the efficacy from the spring vaccines may wear off before the

Learning to recognize the signs of good health is an important first step in being able to detect when your horse is ill. A healthy horse should be alert, have a good appetite, and seem bright and energetic.

The average rectal temperature for an adult horse is between 99.5 and 100.5 degrees, with temperature readings over 101.5 degrees being considered a fever.

The heart rate of a resting adult horse can range from 25 to 45 beats per minute Find out what's normal for your horse by checking his temperature and heart rate from time to time, and become familiar with his normal readings.

There is a very simple way to check for dehydration. Pinch your fingers on the skin on the horse's neck or eyelid, then let it go. If the pinch quickly disappears, your horse is properly hydrated. If the pinched part remains up or is slow to return to the normal position, your horse may be suffering from dehydration.

You can check the capillary refill time of your horse's gums if you suspect he is in shock. Place your finger on his gums, press for a few seconds until the place you've pressed has turned white, then release your finger and wait for the color to return. The gums should be pink again in under two seconds in a normal horse. If it takes longer than this for the color to return, your horse may be in shock. You can also check for dehydration when you are checking the capillary refill time, if your horse's gums are moist and slimy, this means he is well hydrated.

If you suspect abdominal pain (colic), you will want to listen to the horse's stomach to check for gut sounds. In a normal horse you will hear occasional gurgling sounds from deep inside his stomach. In a colicky horse, the gut sounds can increase and sound quite noisy. On the other hand, in a severe case of colic caused by an impaction, the gut sounds can completely cease. This isn't normal and veterinary help should be sought.

It's important to monitor the water intake of your horse. He should be drinking several gallons per day to maintain good health. Check your horse's water buckets regularly. If you have to fill them often, then he is drinking a good amount.

As you clean his stall, observe the consistency of his manure. It shouldn't be hard and dry, but it shouldn't be sloppy and wet without shape. When you clean the stall, note the number of piles. The average horse produces approximately eight piles of manure each day, so if you are seeing far more (or far less) than this number, this could be cause for concern.

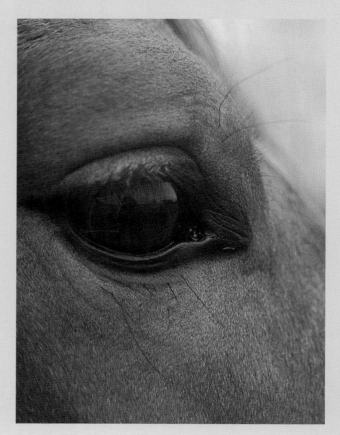

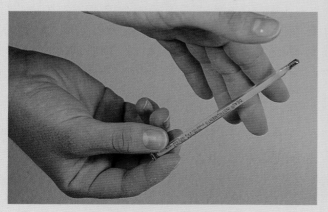

A fly sheet may be necessary to keep the bugs off. When used, fly sheets should fit properly without being too tight and without leaving loose straps that the horse could get tangled in.

helpful if you can have cross breezes blowing) and keep fans running in your barn aisles. Keeping your barn aisles dark will discourage biting flies from entering your barn. Clean the stalls promptly and remove any manure as quickly as possible to prevent the attraction of flies.

One of the safest, easiest, and most inexpensive methods of fly control are sticky fly traps. These come in a variety of shapes and sizes, but we have personally had the best success with the long, thin strips that come in little tubes. The tubes are unrolled and hang from the ceiling. The success rate of flies caught is phenomenal. The more flies that get stuck on the trap, the more new flies you catch. These work exceptionally well at reducing the number of flies in the barn and stalls. Hang them on the edges of the doorways to catch the flies as they try to enter the barn and hang strips in front of each horse's stall. A major disadvantage to these sticky fly traps is that they are

incredibly ugly, plus you'll be amazed by how many times you'll walk into them and get them caught in your hair!

For a horse who is particularly agitated by flies or a horse who has an allergic reaction to flies, you may want to invest in a fly mask or fly sheet to offer some protection against the bothersome creatures.

If your horse is being annoyed by the flies and if fly spray and sheets don't seem to be helping, you will want to bring him inside if possible. Put your horse in his stall, turn out the lights, and get a fan going. The cool, dark atmosphere will help keep the flies away.

LICE

Lice are often more troublesome during the winter months and burrow down inside the horse's deep winter coat. The type of lice that affects horses is not the same type that affects other animals or humans and can only be passed

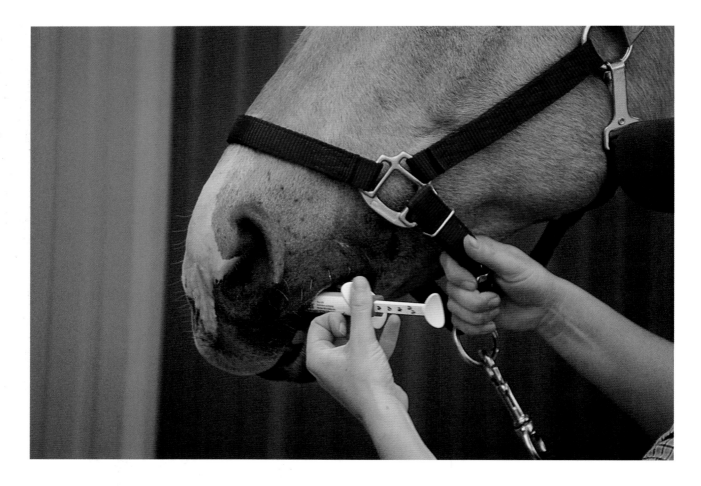

Deworming a horse is not a difficult task if he has been properly introduced to the procedure. Remember to try and raise his head until he has swallowed the paste to prevent him from spitting some back out, thus under-dosing himself.

from one horse to another. Unfortunately, they do this with great ease! Consult your veterinarian regarding the proper treatment for your horse. It may involve more than one session in order to be certain that all of the lice are eradicated. Additionally, any equipment used by the infested horse must be thoroughly cleaned. This includes blankets, brushes, and leg wraps. Ivermectin is also effective in treating lice.

MOSQUITOES

Mosquitoes are of particular annoyance to horses and have the added irritation of biting. Some horses are allergic to mosquito bites and anemia can result if a horse is continually exposed to multitudes of mosquitoes. In addition to all of this, mosquitoes also carry the Eastern, Western, and

Venezuelan varieties of equine encephalomyelitis, as well as spread West Nile virus. Mosquitoes are at their worst around sunset so you may want to consider having your horses stabled in the barn at this time of day. Mosquitoes are typically less bothersome inside the barn. Repellents and insecticides are also helpful in controlling the annoyance of mosquitoes, but you must be careful when using these, as you don't want to irritate your horse's eyes. It's also important to take caution to avoid getting any of these products in your horse's water.

TICKS

Ticks are troublesome insects that feed on blood. It is important to remove them when you see them, and doing a quick visual check of your horse when you bring him in from pasture can help you detect and remove the ticks before they have attached themselves to your horse. Remove any ticks that you see, but you should not do this with your bare hands. It's best to wear rubber gloves when removing ticks. They can carry diseases that are transmissible to humans, including Lyme disease, anaplasmosis, and Rocky Mountain spotted fever. Regular grooming will help you notice any ticks that may be lurking in your horse's coat. Pay special attention to his face and neck, as well as his armpits, chest, legs and the dock of his tail. Ticks can be treated with an insecticide containing pyrethrin.

INTERNAL PARASITES

A proper deworming program to protect your horse against internal parasites (worms) is vital to your horse's health and well-being. You will want to talk with your veterinarian for recommendations to give you exact specifications on the deworming program that will be best suited to your horse's needs. Your deworming schedule will vary depending upon a number of factors, including your location, whether your horse is kept at pasture or in a stable, and your horse's age.

There are several types of worms that affect horses, including:

 Strongyles (Large=bloodworms, small=cyathostomes)
 Roundworms (ascarids)
 Pinworms (oxyuris equi)
 Bots (gasterophilus)
 Tapeworms (anoplocephala)

Due to the fact that there are different types of worms, your deworming program will likely include different types of dewormers. Some of the most common types of dewormers are:

 Avermectins (Ivermectin and Moxidectin)
 Pyrantel (sold under the brand name Strongid, among others)
 Benzimidazoles (sold under the brand names of Safeguard and Panacur, among others)

Most horses are kept on a rotational deworming program, which involves alternating between the various products. Many veterinarians will recommend the use of

A horse with a gleaming eye and shining coat is generally one who is in excellent overall health and condition. A healthy horse is alert and attractive.

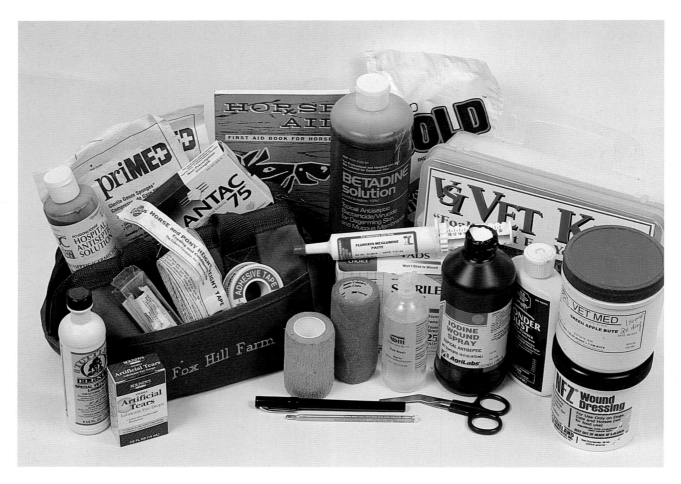

This sample equine first-aid kit demonstrates some of the items you may want to include in your own kit. Products that are not frequently used should be routinely removed and replaced.

an ivermectin product at least twice per year, typically in the spring and in the fall. It is usually recommended that you deworm your horse with a paste deworming product (squirted into the mouth) every 6 to 8 weeks, with variations depending on your horse's personal needs. There are also daily dewormers that come in pelleted form and can be put in your horse's feed once a day. These products provide daily protection against some types of worms, although you will probably still want to supplement with an ivermectin product in paste form at least once a year. Some worms may develop resistance to the daily dewormer and the product may prove to be less effective after long-term use.

Paste dewormers come in tubes that are inserted into your horse's mouth. Because paste dewormers are measured by your horse's weight, make sure that you double check the dose you are giving to ensure that you do not overdose the horse.

FIRST AID

While it's very important to know the basics of first aid so that you can help your horse in the event of an injury, in most cases you will still want to put a call in to your veterinarian. Even if the vet doesn't feel that the injury warrants an emergency visit, he or she can still give you expert guidance and insight into the correct treatment protocol for your horse's specific needs. Plus, you will have the comfort of knowing that you are following the proper procedure in treating the injury.

There are many books that detail the process of cleaning and medicating wounds, wrapping bandages, and the proper treatments for various types of injuries. Having

A horse owner demonstrates some simple first aid of washing a wound and running cold water over it to reduce swelling. It's very important for horse owners to understand the basics of first aid.

After suffering a stumble and striking her leg, this filly was taken to the veterinarian for proper bandaging. If you are at all unsure about the proper procedure in a certain instance, it's always best to get a veterinarian's advice.

a few of these books in your home library are well worth the investment in the event that you are able to assist your horse during an emergency.

COMMON AILMENTS

As you take care of your horse and spend time with him each day, you'll know immediately if something is not right. It may be that his eye is red and oozing with discharge or he may seem a little uncomfortable and not quite as interested in his food. Learning to recognize the symptoms of common equine ailments will be helpful to you in discerning what is normal, what isn't, and when you should call the vet.

Choke is a blockage of food—usually grain—in the esophagus. Symptoms are characterized by coughing,

Stocking your First-Aid Kit

What are the important things to have on hand for your first-aid kit? Buying a pre-packaged vet kit or first-aid kit from a veterinary catalog can give you a good jumpstart on accumulating the proper products to keep in your barn or horse trailer. Or you can start from scratch and gather precisely what you feel that you would like to have in your kit. Some good items to get you started are:

Antibiotic ointment	Scissors
Bandages	Sterile dressing
Betadine	Stethoscope
Chlorhexidine	Surgical tape
Cotton balls	Thermometer
Disposable syringes	Vetwrap
Peroxide	Sheet Cotton
Petroleum jelly/Vaseline	

You may also want to keep a twitch in your first-aid kit, as well as a flashlight with extra batteries. Keep your vet's phone number handy. It's always wise to be certain that your treatment protocol is correct.

Ask your veterinarian about the possibility of keeping Banamine or phenylbutazone (bute) in your first-aid kit, as well as getting information on the proper dosing instructions.

unwillingness to eat, and liquid with food particles dripping from the nostrils. Do not feed any additional food or water, as this will only increase the blockage. Treatment can include tubing with mineral oil (performed by a veterinarian) or massaging the neck. Since choke often occurs in horses who bolt their grain (rapidly consume large bites without enough chewing), this can often be prevented by feeding grain in smaller amounts more often. Or you can place a large stone in the bottom of their feed buckets, which prevents them from being able to gulp such large quantities of grain at once.

Colic is abdominal pain caused by digestive upset, such as gas or impaction. Symptoms can include frequent rolling, loss of appetite and unwillingness to eat, tail swishing, kicking at the stomach, biting their sides, sweating, pacing, or a listlessness. Colic can be caused by numerous factors, including a change in diet or routine,

lack of water intake, too much rich grass, or stress. Treatment options will vary depending on the severity of the bout of colic. Painkillers, such as Banamine, can often relieve the pain and aid in the speed of recovery. Hand walking can be very beneficial in helping to relieve colicky symptoms. Tubing with mineral oil is also very helpful. In extreme cases, surgery might be necessary, but it is an extremely risky and expensive option of treatment.

You can lessen your horse's chances of colic by making all changes to diet slowly and over the period of several days. The same is true for turnout on grass. Make the transition slow and don't allow your horse to have access to large quantities of grass if he is not used to it. Try to insure that your horse is drinking well and fully hydrated by keeping fresh water in front of him at all times, as well as giving him access to salt. Try to minimize the stressful events in his life, particularly if he is an easily upset or nervous individual. Some horses don't mind any changes and even seem to thrive on them, while other horses do not do well if their situations and routine are changed. We have seen horses colic when they were moved to a different pasture not because of a change in the amount of grass, but because they were so upset about being put in a new place. If you know your horse has a nervous tendency, make any changes slowly and try not to make them often. If you can keep things the same for him, you should.

Other ailments include laminitis (inflammation of the hoof laminae, a cause of lameness and a very serious condition), hives (skin irritation or allergic reaction resulting in itchy bumps across the body), rain rot (skin infection caused by a fungus from having a continually wet or damp coat), thrush (a common bacterial infection of the hoof's frog, typically caused by too much exposure to wet and muddy conditions), and ulcers, which are often caused by

Horses who have suffered from the hoof deterioration of laminitis will frequently lay down to take the pressure off of their feet. The condition is often treatable, but not particularly curable.

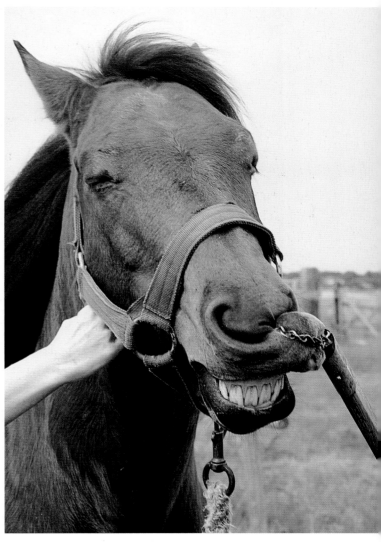

Subduing a horse with a nose twitch may be necessary to accomplish certain veterinary tasks if the animal becomes uncooperative. Using it releases endorphins into the bloodstream, which have a calming effect on most horses.

Repetitive rolling for no apparent reason may be a sign of colic and should be investigated. However, it must be noted that horses roll for many other reasons, such as this mare who was just turned out and is enjoying herself. It is the repetition of frequent rolling over and over that should be investigated.

stress, and are particularly common in foals. Symptoms can include teeth grinding, lack or appetite, or abdominal discomfort. Ulcers are often treated with Zantaz (ranitidine) or GastroGard (omeprozale). In addition, there are always the minor injuries that can occur, such as scrapes, cuts, or a piece of hay poked in an eye.

You will want to discuss treatment options with your vet, as he or she will be able to make the proper diagnosis and recommend exactly the right treatment for your horse's particular condition.

EXERCISE
HOW MUCH EXERCISE DOES MY HORSE NEED?
To insure optimum health your horse should have the benefit of exercise each day. For most horses, this is as simple as being turned out to pasture, walking and grazing and enjoying life with the option of playing, bucking, and galloping if so desired. Horses that stand around (either in stalls or in small paddocks) for most of their days lose muscle, suffer from boredom, and may develop stable vices, such as cribbing or weaving.

Ideally, your horse will be turned out in a good-sized pasture for at least several hours a day, if he is not out all the time. Turnout is extremely beneficial for a horse's mental state, in addition to its physical benefits. The change of scenery, the fresh air, and the chance to run and play all are very good things for horses, in addition to the numerous dietary benefits reaped from grazing.

Obviously, a performance horse in heavy training or use will need more exercise to maintain muscle and fitness than a broodmare who spends her life at pasture raising babies. A performance horse may need exercise each day,

Horses kept in large pastures will have the freedom to run and play at will, a situation that is beneficial to their own health, as well as a beautiful sight for the owner of the horses.

longeing or working under saddle to keep up its condition and fitness.

CONDITIONING YOUR HORSE OVER TIME

Let's say that you just purchased a 10-year-old gelding. He was ridden regularly in a lesson program for five years and was turned out to pasture for two years when his owner lost interest. Now you've purchased him and you're ready to get started on having fun together in the show ring.

You must understand that his long layoff has resulted in an astounding loss of muscle, conditioning, and fitness. He's probably overweight and not used to any sort of physical exertion. Therefore, you won't be able to begin riding

him for hours every day until you have had a chance to slowly build up his stamina. Begin by working him for a short time (half an hour) at the walk, keeping your lessons short and simple. After a couple of days, you can begin to increase the time spent working and perhaps introduce some short periods of trotting. The important thing to remember is to keep the sessions short and not to use the faster gaits for long periods of time until he has had an opportunity to build up some muscle and endurance.

Continue increasing his working time until he is working comfortably without undue stress. You don't want a tired, sweaty horse whose muscles are sore from overexertion. Take things slow and give him time. Once you have

If your horse is unfit and you are slowly building his endurance, it may be a more enjoyable choice for you to ride him quietly on the trails than going around the arena for long periods of time at a slow gait.

built up his stamina and have him working well for longer periods of time, you can begin to consider taking him out to shows or heading out for longer trail rides.

HOOF CARE
FINDING A FARRIER

Finding a good farrier for your horse is an important task because you are entrusting the care of your horse's hooves to this person's capabilities. You want to find a farrier that is knowledgeable, experienced, trustworthy, and punctual. The latter characteristic is probably more important than you might realize, but spending a few afternoons waiting for your farrier to show up for an appointment is enough to make you place this item high on your list of needs.

Ask around for recommendations when you're searching for a good farrier. You should be able to get a list of names by asking barn managers, trainers, tack shop or feed-store owners, other horse owners, and your veterinarian. These people will be able to recommend skilled farriers with expertise and training and can steer you away from farriers with less than positive reputations.

Once you've found a good farrier you will want to treat him with courtesy. This means not being late when you have an appointment scheduled (you can't expect your farrier to begin without you), having your horses ready (already caught and ready to go, not who-knows-where out in the back forty), and paying him in full when the appointment is finished. It's unfair to the farrier to make

Consistency is the key to conditioning. If you are unable to devote the time to ride your horse on a particular day, it may be a good idea to take him out and give him a session on the lunge-line, a task that will take you less time to perform while still maintaining his exercise program.

him wait while you catch your horses. His time is valuable and wasted time at your farm can ruin his schedule for the entire day. The same is true for appointments with your vet. Be mindful of being prepared and keeping the appointment running smoothly. If it's a hot day, offer a cold drink. Working with horses in the heat (whether it's trimming hooves or floating teeth) is hard work and refreshments are always appreciated.

SHOULD MY HORSE BE SHOD?

For decades it was taken for granted that if a horse was doing anything beyond standing in a field, he needed shoes to protect his hooves. Today, while the majority of working horses (those being ridden or driven) are still shod, it is becoming more common to leave horses barefoot. Some farriers feel that more horses are shod than necessary and that many horses thrive when left barefoot, even when being worked.

Shoeing is probably necessary if your horse is being worked heavily on hard surfaces (concrete, pavement, rough terrain) or if he is predisposed to having weaker hooves that may need better protection.

If you and your farrier determine that your horse should be shod, you will want to schedule farrier appointments every 6 to 8 weeks. At each appointment, the shoes will be removed and the hooves trimmed. Then, the

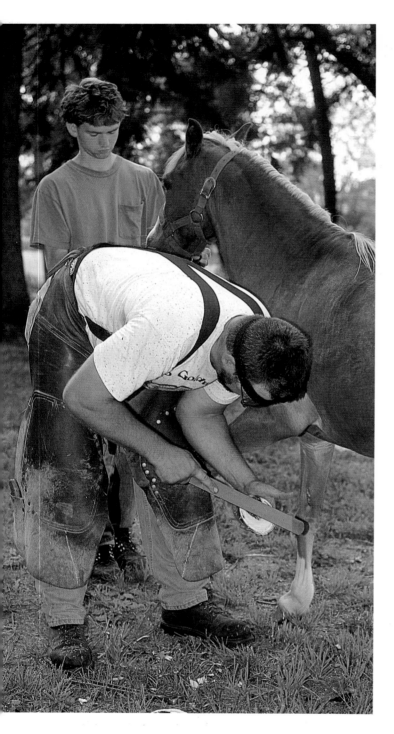

No hoof, no horse! The services of your farrier will be invaluable to retaining the long-term health, performance, and possible resale value of your horse. One of the first things a prospective buyer will look at is the hoof condition. If you ever decide to sell your horse, the routine care of a farrier will have been well worth the cost.

Picking out hooves on a daily basis will not only keep his feet healthy, it will also double as ground training for your horse and make him easier and safer to work with on a general basis.

original shoes will either be reused or be replaced with new shoes. Some horses will only need shoes on the front feet. Shoes on the hind feet are considerably more dangerous because if the horse kicks another horse or a human, the impact can be disastrous if the hoof is shod. This is one reason why you must carefully consider the pros and cons of shoeing, especially if children will be around your horse.

REGULAR TRIMMING

Whether your horse is shod or barefoot, his hooves will need to be regularly trimmed. Many horse owners will say that regular trimming is unnecessary as their hooves will wear off by themselves just by regular exercise. Others will point out "so-and-so old horse" that has never been trimmed in his life and has never taken a lame step. Or they may reference the wild horses who live for years out in the mountains and are never touched by human hands. They aren't being trimmed every 6 to 8 weeks, so why should yours?

Hooves grow at the rate of approximately 1/4 inch per month, so as you can imagine, it doesn't take long for hooves to grow to a length that can cause damage. Some of this growth is worn off by natural wear as the horse exercises. However, hooves generally grow faster than a horse can wear them down and thus, trimming is required. Trimming also prevents the hooves from becoming unbalanced as they grow and helps keep the proper pastern angle. Horses wearing shoes are unable to wear down hoof growth because their hooves are protected by the shoes, and therefore, they must be trimmed regularly.

You will want to schedule regular appointments with your farrier for trimming, whether your horses are shod or barefoot. These appointments should probably be arranged for 6 to 8 week intervals unless there is a specific reason for your horse to be trimmed more or less often. Special attention should be given to foals' feet as their hooves can grow very quickly. If their hooves are left too long, the conformation of their legs may be altered as the foals grow. Be certain that you pay close attention to the condition of your foals' feet as you may need to have your farrier out more often if they are growing rapidly.

Hooves grow in the winter, despite what you may hear from some horse owners. Granted the rate of growth in

It may be necessary to stop and examine a particular hoof during your ride if you suspect some sort of problem. A tiny rock or other foreign object wedged near his frog is an uncomfortable sensation to your horse. Shoes are manufactured out of different materials in a wide assortment of varieties. Some are nailed on traditionally and some use glue. Your farrier will know which method will best suit your horse and situation.

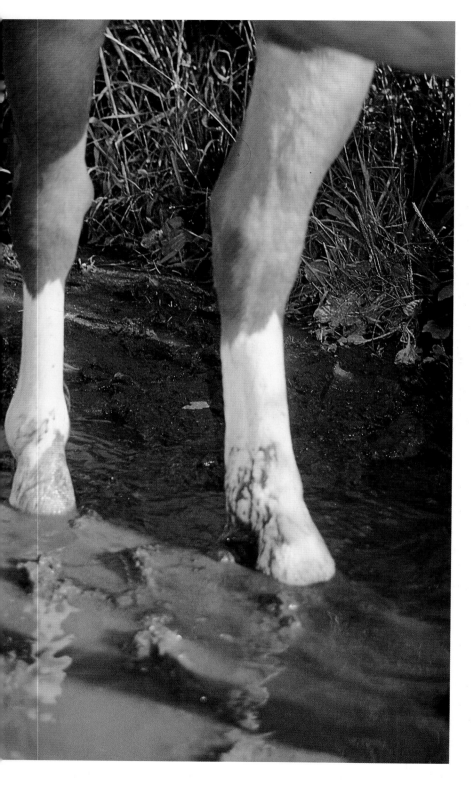

Horses who are forced to stand in muddy paddocks for hours are more likely to develop hoof problems, such as thrush, even despite regular cleaning.

winter may be somewhat slower than the rate of growth in the summer, but they do still grow. You will want to schedule farrier appointments during the winter to maintain the proper care of your horse's feet.

PICKING OUT HOOVES

As a part of your daily care routine, you will want to take the time to pick out your horse's hooves. This daily check will alert you to any problems that may be brewing in the hooves, as well as allow you to remove any object that might be lodged within the hoof, such as a stick or a rock. Picking out the hooves removes any packed mud or manure and helps prevent thrush from occurring. You will also want to pick out the hooves after you've been riding, especially if you've been out on the trails where your horse might be more likely to pick up an object in one of its hooves. Daily examination of the feet is also important if your horse is shod, as you will want to take a quick check of the shoes to ensure that none have come loose and that the nails have not shifted.

COMMON HOOF PROBLEMS

Thrush is a bacterial infection of the frog that often occurs when horses are kept in wet or muddy conditions. The frog begins to rot and the hoof begins to smell. There is a distinctive black tarry substance around the frog. This condition may result in lameness if left too long without treatment. Therefore, you will want to check your horse's hooves often, especially in the winter and/or if you're having trouble with muddy paddocks. If there is any suspicion of thrush, the hoof should be picked out and the inside should be rinsed with a diluted (1:10) bleach solution or a 1-percent

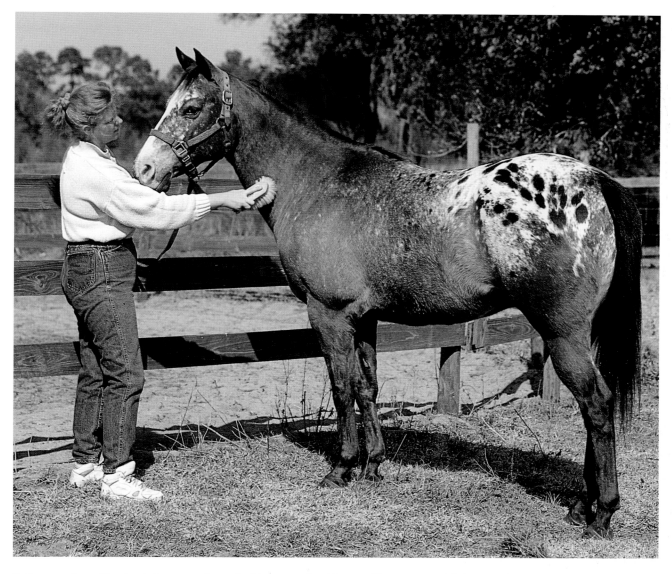

Daily grooming will not only keep your horse looking great, but it's a terrific way to establish your relationship with him. It is also a good chance to look him over carefully and keep a watch for any possible injuries.

tincture of iodine. Consult your veterinarian regarding his or her recommendations and duration of treatment. Obviously prevention is the best method and thrush can be avoided by keeping your horse from standing in wet conditions, frequently picking out his hooves, and keeping a watchful eye for any signs of thrush.

An abscess occurs when an infection enters the hoof. The pocket of infection must be drained, thoroughly cleaned, and covered. Your veterinarian may prescribe soaking in warm water with Epsom salts and/or a round of antibiotics.

Laminitis is a very serious condition and is a cause of lameness. While many people automatically associate laminitis (founder) with ponies, it must be understood that horses can be affected by laminitis as well. Laminitis is an inflammation of the hoof laminae. This inflammation results in extreme pain for the horse. If the laminitis continues it may result in rotation of the coffin bone. Laminitis can be caused by many things including overeating, stress, access to too much rich pasture, difficulty during foaling, and too much concussion of the hoof on

A clean, well-stocked grooming kit will be the envy of the barn! Everyone wants to have a grooming box this neat, tidy, and complete. Be on the lookout for trespassers who will want to borrow from you!

hard surfaces. Initial treatment can include phenylbutazone and special shoeing. However, it is important to determine the underlying cause of the laminitis. Some horses suffer from medical conditions such as Cushing's disease or insulin resistance, which can make them more prone to bouts of laminitis. Warning signs of laminitis can include lameness (pain in the hooves), a digital pulse, warmth to the feet, resistance to walk or trot, and difficulty turning.

GROOMING

Grooming is a wonderful win/win situation. As you brush and groom your horse, you are spending time with him and paying attention to him. You're giving him a thorough examination for any cuts, bumps, bruises, or bites. You're also teaching him about patience and the importance of standing still and making him look great in the process!

Ideally you will devote some time each day to grooming your horse, though you may not have time to do a complete and thorough job each and every day. However, it's important to take some time each day to look over your horse, check for any problems, remove any burrs from his mane and tail, wipe off any mud that is caked on his coat, and generally examine him to ascertain good health and condition.

To properly groom your horse, you will want to have a variety of brushes available to ensure that you are able to do a thorough job. While not technically a brush, a rubber

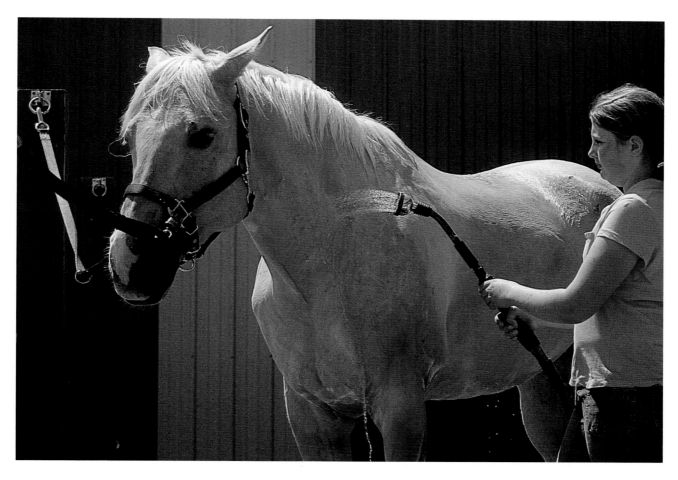

A light shower adapter added on to the end of your hose will not only give you control over the strength of the spray but will be a less-intense pressure and more comfortable to your horse.

curry comb is the best grooming tool to begin your grooming job. Grasp it in your hand and move it in a circular motion, which will loosen up the dirt and bring it to the surface of his coat. Follow this up with a stiff body brush and use swift, sweeping strokes to remove the loosened dirt. Then you can use the softer dandy brush to finish up the job. A mane and tail comb are suitable for working through any tangles in the mane and tail. A regular human hairbrush will also work very well and is sometimes easier to maneuver than a mane and tail comb.

PREPARING A GROOMING KIT

While some experts will recommend that you keep a separate grooming kit for each horse, this might not be feasible for you if you have several horses. If you are boarding your horse(s) at a stable, you will obviously want your own grooming kit and won't want to share these items with other horses at the stable in order to prevent the spread of diseases, skin conditions, or parasites such as lice. If you keep your horses at home and have one with a skin condition, you will want to keep his grooming equipment separate and only use those items on him. But if you have a few horses and they are all healthy and go out to pasture together, there should be no reason why you cannot use the same grooming items for all of them.

Your basic grooming kit will probably include the following items:

Body brush	Rubber curry comb
Cactus cloth	Rubber grooming mitt
Dandy brush	Scissors

Types of Grooming Tools and Brushes

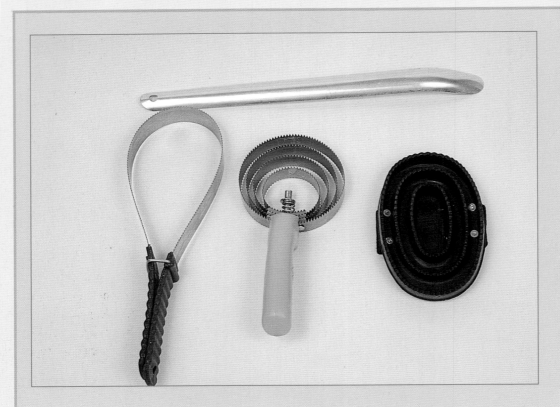

Grooming tools, from left to right: sweat scraper/shedding blade, metal curry comb, rubber curry comb.

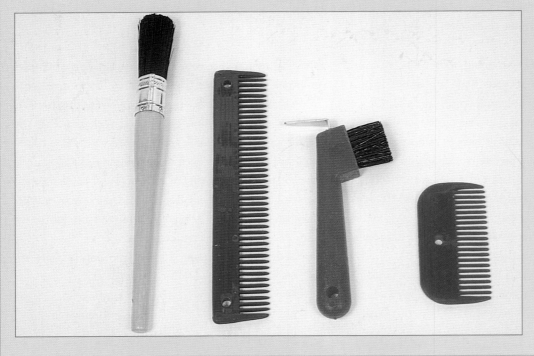

Grooming tools, from left to right: hoof oil applicator, large mane and tail comb, hoof pick, small comb.

An assortment of brushes, from left to right: face brush, dandy brush, body brush, cactus cloth.

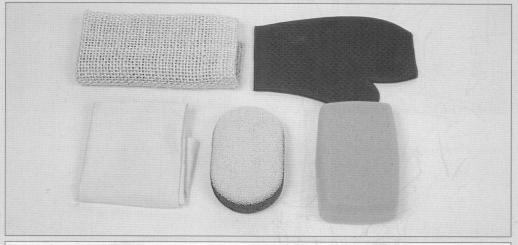

A variety of other brushes, from left to right: long bristle body brush, grooming brush, rice root brush.

Grooming tools, from left to right: cactus cloth, curry mitt, grooming cloth, sponges.

Almost done! After a final rinse, apply a sweatscraper or the flat side of a shedding blade to remove all excess water from the fleshy parts of his body.

Hoof oil
Hoof pick
Human hairbrush
Metal curry comb

Shampoo
Sponges
Sweat scraper

BATHING

If you live in a northern region with cold winters, spring can be one of the best times of the year. One reason is that you can finally give your horse the bath he probably needs. Even regular grooming throughout the winter will not be able to penetrate months of mud and grime. A bath will always leave a horse with a shine that you can't get from grooming alone. During the rest of the year, you will want to bathe your horse on a regular basis for the ease of grooming, particularly if you will be attending any shows with him.

Bathe him on a warm, sunny day—without a breeze, if possible—so he will not become chilled. It is best to tie up your horse to bathe him, either in a wash rack or wash stall. Rubber mats laid over concrete will make the best footing as they will keep the floor from becoming too slippery.

Start by introducing the horse to the hose. Some horses who have not been acquainted with it may initially be afraid or spooked by the hose. Start by lightly spraying his front feet and then work up his legs to the chest. Avoid spraying water near his face and eyes or ears, as horses usually hate that. Later, you can wash his face by hand with a damp cloth or soft face brush. Work your way around his body, starting at his neck and working back, and get everything wet. Be especially aware when spraying near his back

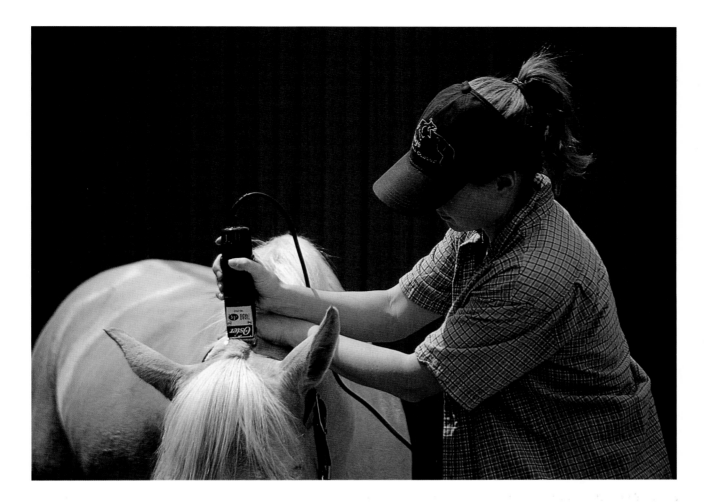

A common clipping practice in some breeds and performance disciplines is to shave a bridle path behind a horse's forelock and a few inches down his neck, removing the mane hairs from that area, and leaving a clean-cut spot for the halter and bridle to rest. If you are showing your horse, be sure to check if bridle paths are allowed and how they should look for the particular event.

legs, as some horses don't like the feeling of drops of water dripping off their bellies onto their legs.

Once he is completely wet, you can apply shampoo suds with a stiff brush all over his body starting with his neck and working back, applying the lather into his coat. You can thoroughly soap up his mane with the brush as well, and include his forelock by pulling it back between his ears with the rest of his mane. You can wash most of his tail by carefully holding the bucket of suds up and dipping his tail into it, and then completing the job by using the brush or a sponge on the top.

Thoroughly rinse your horse in the same fashion as before. Rinse him several times so that you don't leave any soap in his coat or on his legs. Shampoo left in a horse's coat can dry out and irritate his skin. When you come to the tail, stand aside as if you were going to comb it, and rinse from there. When he is fully rinsed, remove any excess water with a sweat scraper, using a firm feel to whisk away the leftover dampness.

Keep a line on him and walk him around some to dry. If you turn him loose or put him in a stall while he's fairly wet he will roll and ruin your nice job. It's terribly frustrating to watch your freshly bathed horse lay down for a roll in his stall! Once he dries a bit you might want to apply a mane and tail or coat conditioner for a final sheen and softness.

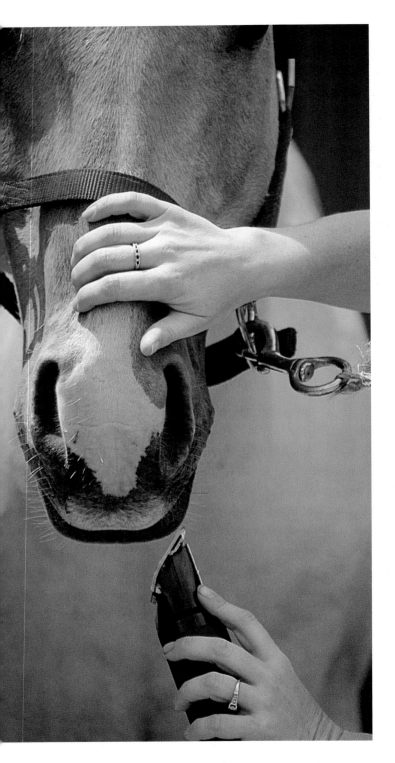

Removing the whiskers from around your horse's muzzle can give a clean, professional look when handled carefully. You will probably want to do this prior to a show.

CLIPPING

If you're showing your horse, chances are you will want to tackle some clipping. It just doesn't seem quite right to head into the show ring with long tufts of hair hanging out of your horse's ears. It doesn't create the image of solid preparation that you're striving to achieve. Similarly, a horse with a beard (long hairs along the jaw) isn't particularly attractive, and neither are 4-inch-long whiskers.

Now, this is not to say that those unattractive hairs aren't useful. The furry tufts in your horse's ears are great barriers for keeping bugs and gnats from bothering him. The long whiskers on his muzzle and around his eyes are good protectors, and if you don't have to remove them, it's probably better not to get rid of them. If your horse is going to spend his summer enjoying himself at pasture, grazing all day and swishing flies, then it's better to leave as much protective hair as possible around these vulnerable locations.

But if the show ring is your aim, then a more polished look is in order. If you're planning on entering in some spring shows and your horse isn't finished shedding out his winter coat, body clipping him is your best bet. This can be a tricky job and it's best to get someone who has a lot of experience to help you.

If you're planning to ride during the winter in an indoor arena, your horse's heavy winter coat may prove to be too much when he is working and sweating. In this case, you may choose to body clip your .Many horses do not like being clipped. The clippers make a noisy, unpleasant sound and many horses are frightened by the sound. Couple this with the fact that some horses are fussy about having their muzzles and ears handled even on the best of days, and you're left with the sort of situation that makes some horsemen run away screaming.

You may have no trouble getting the feathers clipped off of his lower legs, and you might even be all right with getting the body clipping done. But if you're having trouble finishing (or starting) your horse's face, you can try a couple of options. You can take a pair of rather blunt children's scissors and manually cut each whisker or hair away. The problem with this is that there is always the danger of poking the scissors in the wrong place, especially if your horse is jumpy. The last thing you want is a pair of scissors

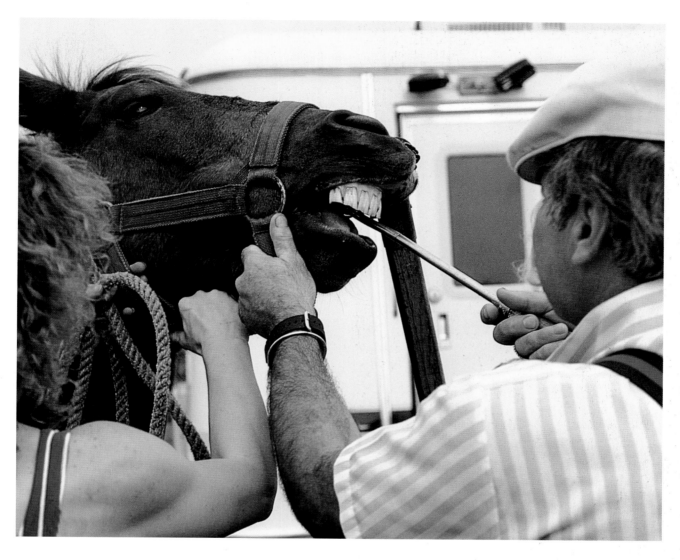

Any sharp edges that have not been naturally worn down by the daily grind of eating will be manually dulled with a file when the veterinarian floats your horse's teeth.

going in his eye. Or you can try a small pair of human nail clippers and snip the hairs away. This is a somewhat safer option than the scissors. Or you can call your vet to give your horse a mild sedative, then finish the whole job without opposition once your horse is sleepy.

DENTAL CARE

A good part of your horse's health is dependent upon his teeth. While many horses go through life without suffering from dental problems, it does occur in a good percentage of horses. Because of this you'll want to make routine dental checks a part of your horse's annual veterinary care. Additionally, you will want to keep your eyes open for any signs of potential teeth problems that might need attention.

SIGNS OF TEETH PROBLEMS

How can you tell if your horse is having problems with his teeth? Certainly, you won't be sticking your hand in his mouth every day to look over his teeth, (even if you could, many horses are resistant to having their mouth and teeth examined), so you will have to rely on the outward signs that can signal when your horse is having teeth troubles.

Puzzling and unexplained weight loss may be a sign of poor teeth. They may not be doing a good enough job at breaking up food and releasing the needed nutrients into the horse.

Many teeth problems can be spotted while the horse is eating grain. If he is continually tossing his head while eating or holding it to the side and up in the air, this may signal he is in pain while he is chewing. Another sign is a sudden slow eater, taking long periods of time to chew grain, and dropping large amounts of it while chewing. This also points toward pain while chewing.

If you notice large amounts of unchewed hay or grain in his manure, this may signify that he isn't chewing thoroughly. Resistance to eating when the horse appears otherwise bright and healthy can also mean that he has some teeth troubles.

Rarely, you may see swelling or puffiness along the sides of a horse's face between the nostrils and the eyes. This can indicate an infection in a tooth and the horse will need veterinary attention right away to get the infection under control. In extreme situations an infected or broken tooth will need to be removed. This may require surgery if the tooth cannot be easily removed by your vet.

Unexplained weight loss can also be a sign of teeth problems, but there are many other causes of weight loss, which are more likely to be the culprit than teeth problems. However, in some cases, when a horse's teeth are causing pain, they will eat less and consume fewer calories. This may result in weight loss, particularly in an older horse.

FLOATING

There are a couple of reasons that you might want to have your vet do a floating on your horse's teeth. Perhaps you want to perform a float as a part of routine health care and

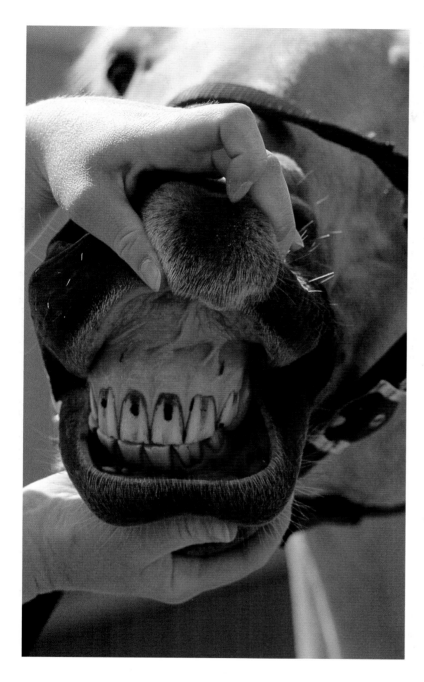

A routine check of the general condition of your horse's teeth performed by a veterinarian can help catch minor dental problems before they escalate.

head off any potential problems before they start. Some horse owners like to do a floating once or twice a year to maintain optimum dental care for their horses by removing any sharp edges on their teeth that can cause discomfort against the insides of their mouths. Some horses are very intolerant of any pain and a routine floating can prevent damages that could interfere with a horse's performance when the bit is in his mouth or when he is eating. Or you may want to float his teeth because you're suspicious of a tooth problem based upon the signs you're seeing.

It's common for young horses (even as young as a year and a half) to need dental attention to help remove caps (remnants of their deciduous teeth) that can get stuck on their permanent teeth as they erupt through the gums. These caps usually come off by themselves and you may find them periodically in your horse's grain bucket after he has eaten. However, in some horses they get hung up on the permanent tooth and cause abrasions from the sharp edges rubbing against the horse's gums and the surrounding area. This can be remedied by a floating from your vet.

In a floating, the horse is sedated and the vet files away at the sharp edges on the horse's teeth. He or she also checks for any cracked or broken teeth, which could potentially become infected. Wolf teeth are sometimes removed because of their location and interference with the bit.

RAISING A FOAL

One plus one equals two. Easy math, but when choosing to breed two horses for the purpose of obtaining a foal, there are many factors to consider in order to produce the best foal possible.

Many people decide to raise a foal without really considering why they want to do so. It's important to consider all of the reasons why you want to raise a foal. Raising a foal is a rewarding and fulfilling experience that many horse owners enjoy, while others would never consider taking on such a commitment and responsibility. It really comes down to whether you are interested in investing your time, money, and efforts in the pursuit of a foal. Breeding a mare and raising a foal are not easy tasks. There are many decisions to make and many months of care and concern. Yet, many horse owners find that it is well worth the effort in exchange for the opportunity to raise a foal of their own and experience the wonder and joy that it brings.

You'll want to consider other factors when making the decision to raise a foal. If you are boarding, can you comfortably afford the additional boarding costs that may be incurred after the arrival of the foal? If you keep your horses at home, does your schedule permit the additional time necessary to care for a foal? Are you comfortable with the idea of your mare being out of commission for a few months at the end of her pregnancy and for the first few months after the foal arrives? Are you prepared to fall completely in love with your new arrival, wanting to spend every waking moment at the barn watching every adorable

move he makes? Is your mare a good candidate for breeding? This includes having correct conformation, being a good example of her breed, having a sensible disposition, and being at a healthy weight for pregnancy. It also includes having a veterinary evaluation performed on your mare to verify that she is in good condition for breeding. If the answers to all of these questions is yes, then you're ready to consider raising a foal.

CHOOSING A STALLION

When you begin researching options for breeding your mare, you may be surprised by the seemingly endless array of stallions available at stud. Pick up any stallion and breeding issue of newsstand horse magazines or start doing some internet searching and you'll find a vast selection of choices. They all seem so handsome, their ads sound as if each one is simply perfect, and you may wonder how you will ever narrow the field down to the right one. Is it even possible to choose?

A quality stallion who has proven himself in the show ring with his exceptional conformation or athletic skill is a good choice to breed to your mare. Finding one who excels in the specific discipline you have planned for the future foal is a wise decision. Or you may wish to save some money on stud fees by choosing a young, unproven stallion who is just beginning his career. Many such stallion owners will often give a financial break to customers to entice them to take a chance with their unproven stock in order to help build a reputation.

It's probably easiest to begin with your plans for the foal. Decide whether you want a purebred foal, or whether you feel that a partbred would be more suitable for your goals. If you decide upon a purebred foal, you've successfully eliminated all stallions except for those of the same breed as your mare. If you want to try cross-breeding, you'll need to determine which breed will be the best match with your mare's breed in order to produce a quality foal with good potential.

Size is another consideration. If your mare is a small horse you might want to try breeding her to a stallion with little more height in hopes of producing a foal that is a little larger. Or perhaps your mare is on the large end of the acceptable height for her breed and you'd like to produce a

foal that is more average in size for the breed. In that case, you'll want to choose a smaller stallion.

Choosing a stallion that complements your mare is one of the most important considerations. Take careful note of your mare's conformational strengths and weaknesses and try to get an idea of what you'd like to improve upon in the foal. Sometimes it can be hard to objectively assess your mare's faults. After all, she is your beloved horse and you love her dearly, but no horse is perfect and it's important not to let barn-blindness prevent you from recognizing the areas where she could use some improvement.

Let's say that your mare has a gorgeous head and neck, wonderful legs, and is a phenomenal mover. In addition, she has a sweet disposition and is a joy to have around the

No horse is perfect. It will be your job to analyze the weak points of your mare and make the best choice of stallion in the effort to produce a foal that will not inherit these faults.

barn. Looking closely at her, you have to admit that she's a little too long in the back and her tailset is lower than it should be. Therefore, your stallion search should be narrowed to stallions that have short, strong backs and correct tailsets, in addition to having good overall conformation, movement, and breed type. In order to carefully plan a breeding with the best possible foal as the result, you'll need to balance the characteristics of both parents in order to achieve that goal.

Your breeding plans may also have an effect on your choice of stallion. If you're planning to breed your mare via artificial insemination (AI), you will need to be certain that the stallion owners offer AI services and find out their specifications and requirements. If you don't want to go the AI route, make sure that you choose a stallion who stands

at stud by live cover. Many stallion owners today offer their stallions by AI only.

Once you've narrowed down your search to two or three candidates, it's time to give them a closer inspection. Whether you take a look by video or in person, you'll want to evaluate each stallion's conformation from all sides, watch how he responds to his handler (Is he well-mannered or pushy and stubborn?), and watch him at liberty to appraise his movement. Ask for information on his show record or inspection scores. These can be good indicators of his overall quality, conformation, and breed type, as well as his talent and athleticism if he has a performance record.

You will also want to learn as much as you can about the foals that the stallion has sired. This is very important

Sending your mare to a stud farm can be a fine option. Here, two pairs of mares and foals graze peacefully on a delightful spring morning to demonstrate the calm serenity your mare can experience in her temporary environment.

because an absolutely gorgeous stallion may not necessarily be siring quality foals. Sometimes the most handsome stallion produces only mediocre foals and an average-looking stallion may outproduce himself with stunning offspring. A stallion who consistently produces offspring with a particular look or similar traits is said to be pre-potent, and if you like the characteristics that he is producing, you can feel somewhat confident in the fact that your foal would have a good chance of inheriting them as well. Try to find out if the stallion's offspring have been shown and what their results have been. If he sires foals that have been inspected by breed registries, find out their overall scores. In what disciplines do his mature offspring excel? Talk to other owners of his foals; ask about their dispositions, good qualities, and experiences. Try to determine which stallion

is most closely producing the type of foal that you seek, but don't forget your mare's contribution to this effort.

Stud fees vary widely in the horse industry, and can range from free (from owners who are introducing young, unproven stallions and want to gain some test breedings to evaluate their offspring), to six- or seven-figure stud fees for Thoroughbred stallions who have excellent racing records. Obviously, these are extremes, and the vast majority of stud fees are under $2,000. When you are setting your budget for breeding your mare, you'll need to figure in more than just the stud fee. Is there a booking fee and is it included in (or in addition to) the stud fee? If you're breeding by artificial insemination, you'll need to find out what additional fees are charged by the breeding farm. What is the charge for each collection and for shipping the semen? Is there a

Artificial insemination can bring the advantages to your mare that a stallion hundreds of miles away has to offer, although this is not always the reason for using AI. If a mare is shy or suspected of being dangerous to breed, such as kicking and showing aggressiveness to the stallion, AI can be used as a safer method of breeding, even if the stallion is on the same property.

For the first five to six months of pregnancy, your mare's routine and daily exercise and work habits should not need to be changed in any way and she can continue to be lightly ridden, driven, or shown as usual.

Your mare will be pregnant for almost an entire year, and if you live in the northern regions of the United States, it will mean having to be extra careful leading her and handling her in icy or slippery conditions to protect both her and her unborn foal from a fall.

deposit due or fee charged for the use of the Equitainer or other shipping container? These additional fees might make a difference in which stallions meet your budget. A lower stud fee may seem attractive, but if each collection is an additional $250 and shipping fees are another $100, it might be sensible not to overlook the stallion with the higher stud fee whose first collection and shipment are included in the stud fee.

BREEDING OPTIONS

Traditionally, if a mare owner wanted to breed his or her mare to a stallion, the situation would involve sending the mare to the stud farm for a month or two before returning home. The advent of artificial insemination has changed this routine somewhat, as it's now possible to breed your

mare to a stallion across the country (or even overseas) without her ever having to leave your hometown. Many mare owners are taking advantage of these opportunities, while other situations still require that the mare be sent directly to the stud farm. A good example being Thoroughbreds, in which artificial insemination is prohibited by the Jockey Club.

SENDING YOUR MARE TO BE BRED

If you've decided to breed your mare via live cover, this will require that she be sent to the stud farm for several weeks. Breeding farms often have specific criteria that must be met prior to the delivery of your mare to their farms. Your mare will need to be up-to-date on her vaccinations and dewormings, have a current negative Coggins test, and have

Almost time! This photograph was taken just two days before this mare gave birth to a delightful colt. All the patient waiting is about to be rewarded.

her hind shoes removed if she is shod. You will want to provide records and information on your mare, including any special feed requirements, habits, or allergies. You'll also want to provide any history of colic or founder, as well as your mare's reproductive history, including whether or not she has foaled previously and the dates of her last heat cycle. Many breeding farms require that your mare be previously examined by a veterinarian and provide a negative uterine culture and cytology to verify that she is in good condition for breeding.

After your mare arrives at the breeding farm, she will be bred on her next heat cycle. She will typically remain at the farm until an ultrasound can be performed to determine pregnancy (at least 14 days post breeding). If she is found to be in foal, then she may return home at that point. However, if she is found to be open, she will need to remain

at the breeding farm to be bred on the following heat cycle, which should follow a few days after the ultrasound (approximately 21 days from the start of her previous cycle). Due to the expense of mare care (a daily fee paid to the stud farm for boarding your mare during her visit), you may or may not want to continue for a third cycle if she is found to be open again. Fortunately, the majority of healthy mares will settle on the first or second cycle.

BREEDING BY ARTIFICIAL INSEMINATION

Location can be a big factor in your choice to pursue artificial insemination (AI) for your mare. Sometimes the ideal match for your mare is too far away for you to feasibly take her to the farm for live cover. Or you may be seeking a rare bloodline that isn't available right around the corner. Perhaps you don't like the idea of sending your

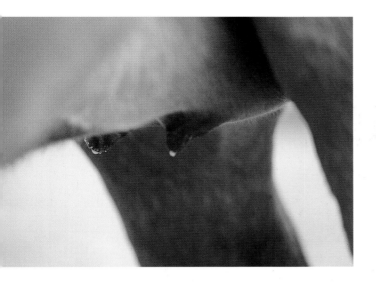

A close-up of dried milk waxing on the udder of a mare who is within hours of birth. Over years of observation, the precise foaling time for a particular mare can be predicted with some accuracy by noting the times of the changes in her milk.

mare away for a month or two and would rather keep her at home where you can be the one controlling her condition and overseeing her care. If she already has a foal at side and you're planning to breed her back for next year, you might not be comfortable with the idea of sending your precious foal off to another farm while it is so young. Whatever your reasoning, artificial insemination may be an option for you.

You will probably be working extensively with your veterinarian or other equine reproduction specialist during the AI process, as it is important to follow all of the protocol very closely to maximize your mare's chances for success in achieving pregnancy. You may find it easier to send your mare to the veterinarian's for a short time while her cycle is monitored and she is bred. It's likely that your vet will manipulate your mare's heat cycles with hormones, by using a product such as Lutalyse before she is bred via AI. Using daily ultrasounds, her heat cycle will

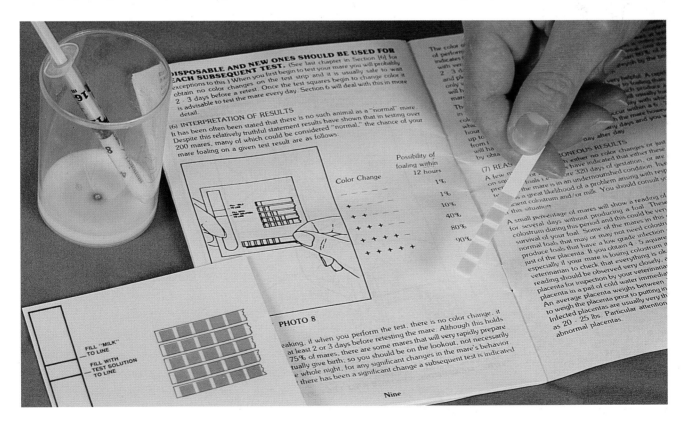

For a maiden mare without an available record to act on, you may want to use a foaling predictor kit, which tests the mare's milk and the changes that occur to suggest how much more time the mare has before foaling.

SUCCESS! A new foal is fast asleep in a soft bed after a good meal of milk. The work and worry of breeding and pregnancy passed, this is the time to reflect and soak in the moment. Foalings should take place on quality straw bedding, deeply applied and free from weeds or other defects. To minimize dampness, some breeders will first lay a light layer of shavings with the straw placed on top of that. The mare and foal should remain in the straw bedding for at least a week.

be documented and semen will be ordered from the breeding farm when the ultrasound shows that ovulation is eminent. After the semen arrives (usually by overnight delivery), the mare is inseminated, and this is often supplemented with a shot of human chorionic gonadotropin (hCG) to induce ovulation.

The next two weeks may seem particularly long, as you wait for the 16- to 18-day ultrasound, which will show whether or not your efforts were successful. The benefit in this case (as opposed to live cover at the breeding farm) is that your mare can return home and be enjoying her daily life while you wait to see if she is pregnant and doesn't need to spend an additional two weeks in the unfamiliar environment. And you won't have to pay the daily board costs, either!

CARE OF THE PREGNANT MARE

Congratulations! Your mare is safely in foal and now it's time to sit back and begin the 11-month wait until your new little miracle arrives. Many first-time breeders find this period nerve-wracking, and face many questions and concerns over the well being of their mare during her pregnancy.

Daily care for a mare in the early months of pregnancy does not differ greatly from care of any horse. Access to fresh water at all times, good quality hay and feed, supplements as necessary, and a rotational deworming program are all important to the health and well being of your pregnant mare. Her nutritional needs will increase in the last trimester of pregnancy and you will need to adjust her daily feed rations to accommodate these changes, but until then she can remain on her pre-pregnancy regimen as long as she stays in good condition.

Many people are concerned about whether or not they can safely continue riding their mare once she is in foal. Light exercise is considered to be very beneficial for a pregnant mare, so continuing to ride her is a good way to keep her in shape. Strenuous exercise is probably not the best idea, and neither is exposing your mare to stressful situations, but light exercise at home through her sixth or seventh month of pregnancy shouldn't be harmful.

You will want to talk with your veterinarian about the possibility of giving your mare rhinopneumonitis shots during her fifth, seventh, and ninth months of pregnancy. You will also want to booster all of her vaccinations approximately one month prior to foaling. This allows the foal to receive optimum antibodies against disease during its first few months of life, as the foal will receive these antibodies in the colostrum (nutrient-rich milk) from his mother.

If you will be moving your mare to a different stall, paddock, or pasture prior to foaling, ideally you will want to get her settled in her "maternity ward" at least two weeks prior to foaling. This will give her time to build up antibodies to her new surroundings. Your mare may be upset at this move and she may resent being removed from her pasture buddies. If she is friends with a particular mare, you might consider leaving them together in the new paddock until the pregnant mare foals, but it's wise to remove the friend after the foal is born. The new mother will only have eyes

What is a Live Foal Guarantee?

Simply speaking, this means that your breeding contract with the breeding farm includes a live foal guarantee (LFG), so that in the unhappy event that your mare does not produce a live foal, you have the option to return for another breeding the following year at no charge. However, this guarantee may be limited by several clauses that could appear in your contract. For instance, the definition of "live foal" can vary. Typically, a live foal is defined as one that stands and nurses. Therefore, if the foal stands and nurses but is lost later due to some unexpected cause, the LFG would be void and you would not be entitled to a repeat breeding. In the case that your foal was aborted or stillborn you will likely be required to provide a veterinary certificate stating this fact. Additionally, some contracts state that the LFG is null and void if the mare does not receive her five-, seven-, and nine-month rhinopneumonitis shots.

Many contracts regulate the period of time in which the LFG is valid. For instance, the mare may return for a rebreeding between May 1st and September 1st of the following year. If the mare owner declines to return her mare for this rebreeding within this specified time frame, the LFG is void. If, by some chance, the stallion of choice is no longer standing at stud, some breeders will allow you to choose another stallion from their farm or they may return your stud fee. Additional fees such as mare care or collection and shipping fees are not included in the LFG and will be your responsibility for the repeat breeding.

for her baby at this point and will no longer want another mare near her. She may become aggressive or overly stressed by trying to protect her foal.

If your mare is going to foal in a stall, ideally this stall will be large and roomy and at least 14 x 14 feet. Always

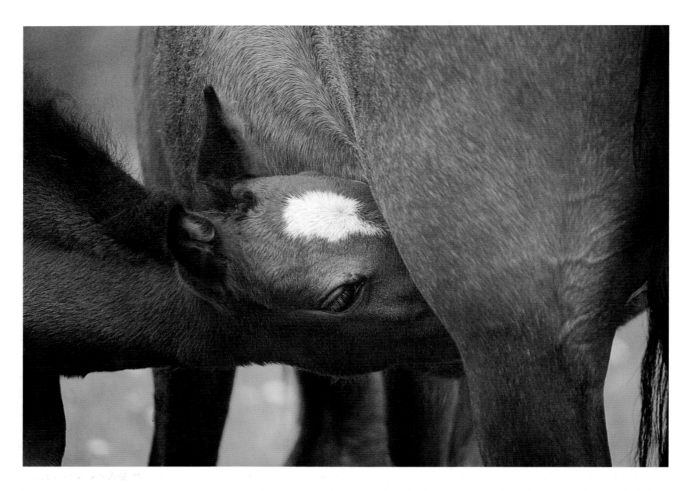

The new foal will nurse frequently during the first few days and never venture far from mother's side. The mare will most likely be extremely protective of her new foal, so you may want to separate them from the other horses for a few days.

make a thorough investigation of your foaling stall before the birth occurs. A stall that seems perfectly safe for an adult horse might have many dangerous possibilities for a wobbly and curious foal. Check for sharp edges and protruding nails or screws. Make sure that there aren't any places where the foal could get caught, such as between the water bucket and the wall. Think carefully over any potential dangers so that your foal arrives in the safest possible environment.

FOALING

The long-awaited magical moment will arrive approximately 340 days after your mare is bred. While 340 days is a good rule of thumb in determining the average length of pregnancy, it is nothing more than a guideline and many mares foal considerably before or after this specific date.

Anything within 320 to 365 days is considered completely normal, and even another 10 days on either end is not necessarily unusual. Studies have shown that mares tend to follow their own pattern when it comes to length of pregnancy (foaling early or carrying longer). Folklore says that colts are carried longer than fillies and that broodmares who have produced numerous foals may have shorter pregnancies than maiden mares. On our farm, neither of these statements have proven true with any consistency.

If you can't rely on dates to tell you for certain when your mare is going to foal, what do you do? Spend 45 nights in the barn "just in case" and promptly lose your mind from lack of sleep? Fear not, as fortunately there are other signs that can give you a better idea of just how close the great event may be.

Twins? Not likely. Foals who are pasture mates enjoy hanging out with each other once they've become friends. Twins in horses are extremely rare, and surviving twins is even rarer. A vet can perform an ultrasound on your mare to determine if she is carrying twins.

If you watch your mare closely during the final few weeks of pregnancy, you'll begin to notice slight changes. The muscles around her tailhead will be relaxing and softening and you may notice a depression on the top of her croup on either side of her tail. This muscle relaxation is in preparation of the foal's birth.

You also should pay very close attention to the condition of your mare's udder. During the weeks leading up to foaling, the udder will grow larger and fuller and change in appearance. Expressing small amounts of the mare's milk can be helpful in determining whether or not foaling is near. This is because the color and consistency of the milk change as foaling nears. Generally speaking, the mare's milk starts out clear and thin, then progresses to amber colored, then amber colored and somewhat sticky, then cloudy white to completely white. This progression can take anywhere from a couple of weeks to a couple of hours. In our experience, many mares will have clear, thin milk for several days and then progress from amber colored to white within the last 24 to 48 hours before foaling. While this is a good guideline, it's important to remember that some mares will deliver the foal while their milk is still clear and the progression to white can be very rapid and may be easily missed if you are not checking frequently. It should also be noted that many mares are very sensitive about having their udders handled at any time, and particularly more so when they are close to delivery, when their udder is swollen and sore. Take great care when checking your mare's milk so that you are not inadvertently kicked, bumped, or pushed.

In addition to the milk changes, your mare may also begin to wax when she is close to delivering her foal. The

Your mare will probably be very protective of her new foal and watch over him carefully and become somewhat aggressive toward her regular handlers. These feelings will pass and within a few weeks she will look out for her baby without such intense emotion.

wax is actually secreted milk that dries on the ends of her teats. The color of the wax will also change in conjunction with the changing of her milk's color, with clear wax being replaced by white wax when foaling is imminent. Typically, it is said that foaling should occur within 12 hours of the appearance of wax, but this is certainly not always the case. We have two mares (full sisters, incidentally) who routinely wax for up to 24 to 36 hours before they finally deliver their foals. We also have other mares who never wax, and we were once surprised by a maiden mare who delivered her foal before she had even begun producing any milk. While the milk changes and the wax are excellent guidelines and certainly are signs that should be closely monitored, it's important to be closely watching for all of the other signs as well.

In the final few hours before foaling, your mare may begin to seem agitated or she may become unfriendly toward her pasture mates. She may graze for a few minutes, then simply stop and stare for a time before returning to grazing. She may kick her belly, swish her tail, bite her sides, or rub her face on her front legs. Some breeders declare that mares will begin yawning prior to delivery. As labor approaches, your mare may appear to be colicking by laying down and rolling, getting up, and sweating, or appear distracted and upset. Some mares go off their feed as well, while others stop eating only long enough to actually deliver the foal.

Once a mare's water has broken, normal delivery proceeds fairly promptly. In a textbook situation, the foal should be delivered within 20 minutes of the breaking of the water. You will see a whitish-looking sack appear from her vulva, followed by the foal's front feet (slightly staggered is normal, it allows for easier delivery of the foal's shoulders). The head will follow, then the shoulders. It is very

Here, a mare and foal enjoy some time together at pasture where the foal is free to explore and learn about the world around him.

important to resist the impulse to help things along and break the sack so that the foal can breathe. Until the shoulders are delivered, it's very possible that the foal may slip back inside, and if the sack is torn, the foal's safety could be compromised. Once the shoulders are delivered, many mares will rest for a moment before the foal's hind legs are delivered. It is only at this point that it is safe to break the sack over the foal's face, and it may be a good idea to do so if the foal doesn't appear to be doing it himself. Once the remainder of the foal is delivered, try to leave mother and baby alone for as long as you can. Be very quiet and try not to disturb them. This allows the foal to remain connected via the umbilical cord for as long as possible, before the cord breaks naturally when the mare stands.

Many foals are standing and nursing within an hour of birth, although it does take some foals a longer time to figure out the coordination of their legs. It can be frustrating to watch the flopping and falling as your foal learns to stand for the first time, but it's good to remember that the stumbling is beneficial in helping to clear out any fluid which may remain in the foal's lungs after birth. Once the foal is standing, the first order of business is for him to find the milk, and most foals begin searching immediately. You may or may not want to intervene at this point. Most foals will eventually reach success on their own and proceed through a sort of trial and error process until they find the udder. On the other hand, if it takes them a long time to find nourishment they can become tired and weak and lose the strength to continue searching. If your foal hasn't nursed within the first two hours, you probably need to help. This may be as straightforward as haltering the mare and holding her still so that the foal can search

When leading a foal who is still being halter-broken, keep the lead rope in your left hand and your right hand free and around his neck near his shoulder. This encourages him to follow you and not pull backward or to the side. It also allows you to hold on and steer him without putting excessive pressure on his neck while he is still young and rapidly growing.

without the disturbance of the mare moving. Many mares (especially maiden mares) are so nervous about their foals that they are only happy if the foal is in front of their face at all times. This makes the situation more difficult because obviously the foal needs to be further back in order to be able to nurse. In these cases, holding the mare can make an enormous difference because once the foal has found the location of the udder a time or two with your help, he will usually able to find it again by himself. By then, he will be strong and persistent enough that he usually succeeds in nursing whether his dam wants to stand still or not.

If holding the mare isn't enough to do the trick, you will need the assistance of another person to guide the foal in the proper direction. We have found that many foals are confused by the necessity of turning their head in order to find the teat, even when they are poking around in the proper place. You may need to assist in order to make sure that your foal is successful in latching on. Then sit back and listen to the wonderful sound of sucking, swallowing, and life-saving colostrum entering the foal's body!

You will need to treat the foal's umbilical cord stump with a 1-percent tincture of iodine or the disinfectant chlorhexidine diacetate (Nolvasan). Do this immediately after birth and then again a couple of times over the next day or so until the cord stump is dry.

Shortly after the foal has nursed for the first time, you will probably see one of two things happen. First, you will

It may be necessary at the beginning to lead a rambunctious foal with both arms around him to keep control without pulling on his halter and neck. The lead rope should only be used as a precautionary measure to keep him from getting away. All other forms of handling and wrangling should be limited to his body.

Foals will attempt to graze very early, trying to experiment and discover what Mom thinks is so great about this grass. Foals mimic many of their dam's behaviors.

probably see the foal pass the meconium (dark, hard stools). If the foal seems to be straining without accomplishing any results, you may need to administer an enema to help things along. Prior to your foal's arrival, you should talk with your vet about the proper procedure for administering the enema or you might want to have your vet do it until you have witnessed how to do it safely and properly.

After the foal has nursed and the meconium is passed, he will probably appear overwhelmed with exhaustion. After all, consider the series of events that he has just encountered! Birth, learning to stand, learning to nurse, iodine on the navel, and maybe even an enema. Don't be surprised if he falls into a deep, sound sleep. After this, he will be up and ready to eat again.

Ideally, the mare should pass the placenta within two to three hours of foaling. The placenta should be examined to make sure that it is intact and that no pieces have been retained in the uterus, which can cause infection. If the placenta has not been passed after a few hours, your vet may need to administer a shot of oxytocin to promote the passing of the placenta.

There are many wonderful books on foaling, caring for a pregnant mare, and caring for a neonatal foal. There is a vast amount of information contained in these books and we highly recommend that you read and research these topics thoroughly in order to ensure the safe delivery of your foal and the well being of your mare.

CARING FOR A FOAL

It may seem hard to believe, but your foal is finally here, safe and sound! All of the months of waiting and the hours of anticipation are over and at last you can begin to enjoy

Preparing a Foaling Kit

A few essential items that you will want to have on hand before the big day arrives

Camera (for taking multitudes of adorable photos!)

Cell phone

Cotton gauze

Disinfectant, such as Nolvasan (chlorhexidine diacetate)

Enema (consult your vet regarding what size enema for your foal)

Garbage bags

Iodine (1 percent) for the foal's umbilical cord stump

Plastic gloves, regular and shoulder length

Scissors

Soap, mild, such as Ivory

Sterile lubricating jelly

String or baling twine

Tail wrap

Towels, soft cotton, as well as paper towels

Veterinarian's phone number

The natural curiosity of foals can be used to your advantage when introducing new and potentially scary objects. Usually curiosity wins over fear and most foals can't resist checking out a new object.

After separation, observe the foal carefully, particularly if his mother is still visible in another field or stall. The foal may investigate the possibilities of forcefully reuniting himself with her.

the fruits of your (and your mare's!) labor. A healthy foal is bright and vigorous, nursing often (every half hour or so), and full of energy, enthusiasm, and spunk.

Much has been written on the subject of imprinting your foal, and there are books on the subject that discuss it in great detail. Essentially, imprinting is desensitizing your foal to various objects and getting him used to people from a very early age. Many breeders feel that it results in positive benefits as the foal matures. Other breeders feel that imprinting interferes with the vital bonding that must occur between mare and foal.

Whether you decide to proceed with imprinting or not, you will need to introduce your foal to basic handling from an early age. You may not have realized it, but when you were helping the foal to nurse or putting iodine on his navel stump or drying him off after birth, you were giving him his first lessons in being handled. Some foals take to humans immediately and will follow you anywhere, expressing the utmost interest in anything you are doing. Other foals will seem overcome with shyness, will dash away when you come near, and will hide behind their dam. These are simply innate parts of their personalities, and yet many first time foal owners may feel as if they've failed when their foal is skittish. Don't worry, it's nothing you've done. You can raise two different foals in exactly the same environment, handle them in exactly the same way, and you can still end up with one that is very affectionate and interactive and one that is less confident.

You can begin to teach your foal the basics of haltering and leading when he is just a few days old. The key to

Many foals can often be seen standing near or under their mother's tails, trying to take advantage of this protection from flies. Foals are often less tolerant of annoying flies than older horses.

remember at this point is to be very careful not to exert too much pressure on the foal's delicate head and neck. Quick and easy lessons are all that is necessary when the foal is very young. A daily lesson of following Mama out to pasture while being haltered and led is probably the easiest way to teach a young foal. By the time he is a few weeks old, your foal will be leading like an old pro.

This brings us to another important topic, which is the fact that foals go through stages. They will test you. They will try naughty things. This is simply part of raising foals. You can halter and lead your foal every single day for weeks, and then one day he will decide that he doesn't really feel like walking along next to you. He may stop and refuse to move forward or he may rush forward at a frantic pace and drag you behind. Or you may deworm your foal every

month without issue and then at four- or five-months-old the foal suddenly acts as though he has never seen a deworming tube in its life. These frustrating little trials will make you scratch your head and wonder what on earth you've done wrong. The truth is you've done nothing wrong. These are just stages that your foal is going through and you need to wait them out. Try not to become upset and calmly continue on and correct the behavior. Keep working with your foal and before long he will have forgotten about being naughty. Foals are smart and will be looking for ways to test your authority to see just who is in charge of the barn. Make sure that the person in charge is you.

Your foal should be dewormed regularly whether he likes it or not. Consult your veterinarian for his or her recommendation as to how often and what products to

A) A healthy, sound, intelligent, mare with good conformation is . . .

B) . . . bred to a quality stallion whose attributes will only improve on hers, whereupon she gives birth to . . .

C) . . . a healthy, active, energetic foal who is a far better example of the ideal equine then either of his parents.

Foals will do odd and seemingly bizarre things, as displayed here. All foals experiment with consuming their mother's manure, which is a surprisingly natural instinct designed to encourage the development of beneficial bacteria in their digestive system.

use. Hoof trimming should begin within the first 6 to 8 weeks, depending on the rate of growth of your foal's hooves. Subsequent trimming every 6 to 8 weeks is important to maintain proper leg conformation and hoof angles during these months when your foal is growing so rapidly. To make your foal's first visits with the farrier a success, you will need to have worked with your foal on having his legs handled and having each one picked up without fuss. If you are fortunate enough as we are to have a farrier that is good with foals and enjoys working with them, then you are very lucky because it will make those first few trimmings much less stressful on your foal.

You might be surprised at how quickly your foal begins mimicking his mother and commences with trying to eat real food. By the time they are a few days old, many foals are already trying to graze and pick up bits of hay or straw. Of course, without teeth they are merely going through the motions of imitation and aren't really chewing or consuming much of anything. By the time they are a couple of weeks old, their front teeth are in and they will begin eating bits of grass and hay. The introduction of grain usually happens very gradually. Curious foals poke around in their dam's feed buckets and pick up bits and pieces of grain. After a few weeks, you can provide your foal with his own feed bucket and a

Selling Your Foal

This is the part of raising a foal that you won't relish thinking about. When your foal is brand new, staggering around the stall for the first time and trying out those spindly legs, it seems impossible to comprehend that you'll ever be able to consider selling him. If you have the room and ability to keep him, you'll probably find great pleasure in raising him, training him, and ending up with a mature, talented horse that you've brought along carefully. On the other hand, you may be limited on the number of horses that you can maintain or you might have planned this entire breeding with the plan of selling the foal after weaning.

If you decide to proceed with the marketing of your foal, you'll want to put as many aspects in your favor as possible. Go ahead and apply for registration papers from any registry that your foal qualifies for. Registration is always a positive asset to any horse or pony, and particularly to a foal. Buyers who are looking at a foal are purchasing on promise, the belief that this particular youngster has the potential to be a high quality adult horse; therefore, pedigree and registration are part of this.

Carefully consider your price. Evaluate the prices of similar foals (breed, sex, size, conformation, color, pedigree) and try to price yours accordingly. While you don't want to price him so high that you never sell him (or maybe you do if you really want to keep him!), you also don't want to underprice yourself and give him away for less than it cost you to raise him. Raising foals costs money and providing high quality care costs even more. There is a value to the efforts you have gone to in providing regular deworming, farrier care, high quality feed and hay, and time in training, so don't underestimate yourself.

Once you have made the sale and your foal is off in his new home, you'll certainly feel a bit lonely for a while without that charming little face in the barn. But you can be comforted in knowing that you provided a happy, healthy beginning to the new life, and you will derive great satisfaction from hearing about how much his new family adores him and how much fun he has brought to their lives, and that will make it all seem worthwhile

The thought of selling your adorable little hoofed wonder may be difficult to accept, especially after the months of time and effort that you have devoted to him.

small portion of grain. This amount can be increased slowly until he is weaned, at which time he should be consuming an appropriate portion of grain and be grazing and/or eating hay as well.

What is the right age to wean? The answers to this question vary widely. In the wild, foals are not weaned until they are approximately a year old, which is usually right before the mare delivers her next foal. Modern day horse owners will wean their foals as early as three months in some cases, while others prefer four or five months. Others will recommend that you wait until at least seven months. Each opinion is backed up by completely plausible reasoning so it's really up to you to decide what you are most comfortable with in regards to your mare and foal. Some owners who are eager to have their mare return to competitive showing may want to wean earlier. If the mare is older and not keeping her weight very well, the owner may decide to wean early in order to allow the mare to improve her body condition. The early weaning of foals with certain orthopedic problems can give you complete control over their dietary needs. But if you are not in any particular hurry to have your mare return to her former career, if she's in good body condition, and if your foal is thriving and healthy, you might opt to leave them together for six or seven months. Some research has shown that foals that are weaned later are less apt to adopt stable vices such as cribbing or weaving.

The possible monetary reimbursement may help soften the blow of selling your foal. The joy that your foal will bring to its new owner is another splendid benefit and makes the parting a little easier.

ENJOYING YOUR HORSE

Sometimes the simple moments of solitude spent enjoying your horse's company are the best. A quiet time spent just watching or observing the beauty of your new animal can be a delightful experience.

Congratulations! You've done it! By now, you've purchased the perfect horse, learned about his personality, developed a feeding program, and decided whether to board or keep your horse at home. You are also providing top-notch health care, and you've considered the possibilities of raising a foal.

Now it's time to have some fun! There are many different things that you and your horse can do together so consider the possibilities and enjoy yourself. Whether you decide on trail riding, dressage, or attending breed shows, you'll have a wonderful time as you explore these new opportunities and learn more about your horse in the process.

LEARNING TO RIDE

You may be thinking, "But I already know how to ride. I rode as a kid. I went to camp and rode horses and I only fell off once." In that case, you may consider yourself already a proficient (or at least, a fairly proficient) rider.

However, as disappointing as it may be to hear, riding a couple of times as a child doesn't really count as knowing how to ride. You may know how to sit on a horse and you may know the "cluck" to go and "whoa" to stop routine, but learning to ride properly takes time and education and, ideally, lessons.

Many riders want to ride for the fun of it; for the thrill of seeing the world from horseback and the enjoyment of hours spent in the saddle. They don't really care whether their position is correct while doing so and don't give much thought to sitting up straight and looking forward. Who cares where your hands are? It's just supposed to be fun, right?

Even if you rode all the time as a child, you will probably want to take lessons to redevelop your riding skills and increase the safety and effectiveness of your abilities.

Correct riding position makes an effective rider, and the more effective you are as a rider, the better you and your horse will be able to communicate. The better you and your horse communicate, the more fun those trail rides and hours in the saddle will be.

Even if you don't have aspirations to show, you might consider a few months of riding lessons just to give yourself the best start as you begin working with and riding your horse. A trainer can be an infinite help in working through any little difficult situation that might arise as you and your horse get to know each other while riding. Your own confidence can grow as you work and learn under the tutelage of

an expert, and you will learn things about riding that may positively effect the time you spend enjoying your horse.

You can take lessons at a riding stable on the stable's school horses, which can be an excellent way for a timid rider to begin. You can also trailer your own horse back and forth for lessons. Many riders prefer the latter situation, as they can utilize their own riding lessons as training sessions together with their horse. Or you can board your horse at the riding/boarding stable (see Chapter 5) and have your horse conveniently accessible for every lesson you take.

In any case, the importance of proper instruction should not be underestimated. You are investing a great deal of time and effort into your horse and his care; so try to make the most of your riding experiences together by making sure that you have mastered the basics of riding and equitation.

TRAINING YOUR HORSE

Now that you have figured out what to do about learning to ride, you need to figure out what to do about any training that your horse might need. You might have purchased a horse that is already trained under saddle, as that is the most appropriate choice for a novice owner. But it's also possible that your horse may need a little reminder course of training to polish him up and refresh his memory, especially if it has been a while since he was worked regularly. Perhaps you fell in love with a fabulous three-year-old mare with excellent conformation, a super pedigree, and a kind and gentle temperament, but not trained at all under saddle. In these cases, what's a horse owner to do?

DOING IT YOURSELF

If you are a more experienced horse owner, training your horse yourself can be a fulfilling and rewarding challenge. After all, who knows your horse better than you do, and who better to undertake the task of training and educating him than you? If you decide to pursue training your own horse, you can benefit from the vast array of books, videos, and clinics available. The amount of information at your fingertips is incredible, and the advice and insight from professional trainers can be invaluable as you begin training your own horse. Whether you're starting fresh with a

However, there are times when you might want to schedule a private lesson if there are specific problems or goals you wish to discuss with the riding instructor. The instructor will certainly be glad to provide assistance.

Riding lessons are often offered in a group setting, which can be less expensive than a one-on-one session. These types of lessons can be very enjoyable, as you will be able to learn and laugh with other riders in the group.

young, green horse or working on a tune-up with an older horse that has already undergone some training, you may find that doing your own training is a wonderful experience. Some horse owners feel that there is nothing more rewarding than going into the show ring with a horse that they have raised from birth, trained themselves, and now are ready to show off to the rest of the horse world.

HIRING A PROFESSIONAL

Professional trainers can be a great choice for many horse owners. You may not feel comfortable starting a totally green horse under saddle or you might be working through some problems that you just can't seem to fix. Or it could be that you just don't have the time to devote to daily training sessions with your horse and you would like to maximize the time that you do have available to ride by simply riding, not by having issues with your still-a-little-green-and-spunky horse. In these cases you might seek the assistance of a professional trainer.

Finding a good trainer might be as easy as utilizing the resident trainer at the barn where you board. You might

Training a green horse begins with the introduction of saddle and bridle long before a rider will attempt to mount up. Longeing a horse will teach him the basics of wearing his tack, while developing a work ethic and introducing him to simple voice commands.

also ask for references from friends who have used different trainers in your area. You can visit different trainers in your area and ask questions about their methods, theories, and fees. Ask about horses that they have trained in the past. Are they competing in shows? Are they winning? An efficient and thorough trainer can get a lot of training accomplished on your horse in 90 days, yet you should be leery of trainers who promise miracles in minutes. Training takes time and repetition, and your green horse shouldn't be walking, trotting, and cantering under saddle the first day. It takes many training sessions for a horse to learn the basics, and it's more important to have the basics firmly established than it is to have a rushed training job. You might feel that the training is not progressing as fast as you think it should be, but you must remember that every horse is different. What one horse takes three weeks to learn, another understands in three days. There's no predicting how your horse will respond to training until it

has begun. Check in regularly with the trainer to see how things are going.

SHOWING

If you're like many horse owners, you'll find the idea of showing to be an appealing one. There's something exciting about loading up your horse into the trailer before dawn and heading off down the road to the nearest show. There's the exhilaration of waiting for your class, the suspense of heading into the ring, the nail-biting moment when the judges have their eyes on you, and the hopeful anticipation of hearing your name announced over the loudspeaker as the winner! There are many opportunities for horse owners to show, so let's take a look at some of the options.

BREED SHOWS

For those who have purebred, registered horses, showing at a breed show can be a very enjoyable event. You're only

Showing a horse you trained yourself can be a major accomplishment and give you far more satisfaction from the knowledge that it was your careful work that brought him to this moment.

If you and your horse are having problems that you seem unsuccessful at sorting out, it's best to seek the advice or services of a more experienced trainer, not only to keep you safe, but also to ensure that the horse does not suffer long-term confusion from poor training methods.

competing against those of the same breed and you have the advantage of showing under judges who have specific knowledge of the breed standard and rules for your particular breed. You don't have to worry about taking your Connemara to an open show and having to show under a judge who breeds Arabians and doesn't have any particular knowledge about Connemaras and their breed standard.

Breed shows often have high point systems with year end and lifetime awards to increase the fun of competition and allow breeders to prove the quality of their breeding stock by accumulating these points and awards.

Depending on the popularity of your breed (and your location), you might have access to several sanctioned breed shows within close proximity of your farm, or you might have to drive for many hours (or days) to reach a competition strictly for your breed. Obviously, the more popular breeds will have more options for showing nationwide than a more unusual breed.

Breed shows usually offer good opportunities to campaign young horses due to the increased emphasis on breeding (halter) classes. Many breeders like to show

Here a Welsh Pony navigates a cone course during a driving competition at a regional breed show. Animals of the same size and general athletic abilities can compete against each other in an enjoyable setting.

yearlings or two-year-olds at halter for a season or two to get them accustomed to bathing, trailering, showing, and the hustle and bustle of the show ring while they are still at a young and impressionable age. Once they are old enough to start in performance, they already have good experience in showing and being accustomed to new situations.

Some breeds also offer classes for crossbred horses (such as for registered Half-Arabians or registered Half-Welsh) at their sanctioned breed shows, which expands the opportunities for showing. The drawback to these types of shows is that they can be somewhat expensive, especially if you must travel long distances to attend them, and thus incur additional expenses such as hotels, gas, and time away from work.

LOCAL OPEN SHOWS

You wake up early on a Saturday morning, load your horse in the trailer, drive for 20 minutes, unload, warm up, show for three hours, load back up, and arrive back home by noon. Total time away from the farm: five hours. Total expenses: less than $50.

Local showing at open shows can be a fun and inexpensive way to have some fun showing your horse. Open shows are typically very inexpensive ($5 to $10 for an entry fee for a class), you don't have to stay overnight, and you don't have enormous gas bills. Most open shows offer a variety of classes to appeal to a wide cross-section of exhibitors so you can show in everything from halter classes to showmanship, Western pleasure, English pleasure, hunter classes, driving classes, speed events, and more all in a single day!

Load up the horses for a day of fun! A locally run open show can offer you and your equine buddies many classes that are enjoyable to attend without spending a long time away from home or breaking the bank.

In addition to the low cost of the open shows, you'll also have the chance to meet many local horse people and make good contacts, which is great if you're looking for another horse to buy or if you're trying to sell one. It's also very good if you're trying to get started with a small breeding program. You can get your farm's horses out and about for the public to see and begin to drum up interest in the type of horses you produce.

Even at the small shows, the competition may still be stiff. There are many people who strive for excellence, and just because it's a local show doesn't mean that the exhibitors aren't serious about their horses and their showing.

One drawback to local open shows is that the judges aren't always able to be totally knowledgeable about every breed or discipline, and might have preferences for or bias against a particular breed. This is just a part of showing at multi-breed and multi-discipline shows. Some judges will love your horse and others might not. That's what keeps showing exciting and keeps exhibitors coming back for more!

DISCIPLINE-RELATED SHOWS

Discipline shows aren't a breed show or an open show. They are specifically targeted for exhibitors and enthusiasts of a particular discipline. There are dressage shows, hunter-jumper shows, and driving events. Whatever your interest, there's probably a show with classes strictly limited to the discipline. These shows offer a chance to exhibit in your specific area of interest or

Local open shows may be easy to attend, but that doesn't mean they're easy to win! Often the quality of the horses and the horsemanship is very high and exciting for both the competitors and the spectators. Here an American Paint Horse works his way toward the camera during a Western pleasure class.

expertise. Again, because these are more specialized shows they offer a wider selection of classes in these specific areas with more options (divisions split by the age of the rider, size of the horse, or level of training) than open shows are able to do. These shows usually have highly trained judges with extensive knowledge of the particular discipline. As with breed shows, these shows can be somewhat more expensive and may require more driving time in order to reach them. There might not be an upper level dressage show right around the corner from your horse, but if showing dressage is your passion, then traveling to a special event might be well worth your time and effort.

HALTER CLASSES

Many people thoroughly enjoy showing their horses in halter classes. They bathe, clip, polish, prepare, and strive to give their horse the best possible advantage when they step into the ring. Halter classes are the equine version of a beauty pageant, with the most beautiful, most correct, best conformed individuals standing at the head of the line, followed in successive order by the remainder of the class. Turnout and manners also count and you can invest a lot of time and effort in the preparations for showing your horse in a halter class, including lessons in standing for inspection and moving out in hand at the trot.

Other people find no interest in showing in halter and prefer to point their energy toward using their horse in performance classes and want to spend less time on primping and polishing than the halter enthusiasts.

Halter classes at breed shows are judged against the written breed standard, thereby comparing each individual to the entire standard of excellence and the judge's mental picture of the ideal or perfect example of the breed. Conformation is also of vital importance, as the judge is looking for the most correct individual in the class.

Typical halter classes begin with walking the horse toward the judge, then trotting away and around the perimeter of the ring. The entrants line up and the judge begins the individual inspection of each entry. The horses

Show Checklist

If you're going to a show, it's going to seem like you're taking far more with you than you could ever possibly use. However, it is incredibly frustrating to arrive at the show grounds only to realize that you've forgotten something vital at home. (Yes, that is experience speaking there. I can recall one show where we drove for an hour and a half, only to discover upon our arrival that we had forgotten to bring along a water bucket for the pony. Of course, we had remembered the jug of water, but we had no way to give it to her. A quick trip to the local hardware store solved the problem, but it was an unnecessary expense of $7.99 that we could have saved if we had paid attention before leaving home.)

Most exhibitors develop a show checklist to help them remember all of the items they wish to bring along. Below is an example of some of the typical items necessary when going to a show. There's no doubt that you will tailor this list to your own personal needs, and you will certainly make additions as you go along. By keeping a master list of the items you usually need, you'll save yourself the annoyance of forgetting something important and you will be able to concentrate on doing your best in the ring!

- **A copy of your horse's registration papers**
- **Bridles**
- **Camera**
- **Clippers**
- **Coggins test results (bring your original, as well as a copy to leave at the show office)**

- **Entry forms**
- **Extra bits**
- **Extra cash (you might want to add classes later)**
- **Extra shavings for the trailer**
- **Film**
- **First aid kit**
- **Fly spray**
- **Girths**
- **Grain (always bring some along, you may need it to encourage your horse to load into the trailer)**
- **Grooming equipment, all brushes and combs**
- **Halters**
- **Hay (bring more than you think you will need)**
- **Hay bag**
- **Hoof oil**
- **Interstate health certificate (if you are showing out of state)**
- **Lawn chairs**
- **Lead ropes**
- **Longe line, longe whip**
- **Saddle pads**
- **Saddles**
- **Show bill or prize list**
- **Sunscreen for yourself**
- **Umbrella**
- **Video camera**
- **Water buckets**
- **Water from home, in jugs**

Hunter-jumper shows remain very popular in many parts of the country and typically offer classes suitable for beginner riders or young horses, all the way up to those at a top level, jumping three- or four-foot obstacles.

may then be asked to move again to give the judge a second look at their movement. If the judge is trying to decide between two entrants, they may be asked to trot out again for the judge to have a final look.

The first- and second-place winners usually come back for a championship class later on to compete against the first- and second-place winners from each halter class of the day. The best two animals of this group are named the Champion and Reserve Champion of the show.

Halter classes are a great way to introduce a young horse to the sights and sounds of the show ring. By the time you're ready to start showing him under saddle, he will have already had the opportunity to get used to the show routine of bathing, clipping, loading, and tying. Many horse owners use halter classes as a good way to begin early training with their yearlings and two-year-olds.

Many people also use halter classes as a way to market and advertise their breeding program. If Top Notch Stallion and Super Duper Mare produce A+ Show Champion, then it helps breeders when they have A+ Show Champion's full sister for sale the next year. Show ring wins in halter help validate the quality of a breeding program. Producing consistent winners in the show ring under a variety of judges and competition illustrates that you are producing young stock that will win anywhere and for anyone. It also helps buyers to feel more comfortable buying a youngster if they know that he has already been shown and placed well under several judges. It shows that his conformation and type have been evaluated by expert professionals and deemed satisfactory.

DRIVING

Move over Chevy and Ford, many horse owners are returning to a more traditional form of driving: their horse. This doesn't mean that horse owners are abandoning their cars, trucks, and SUVs and heading to the grocery store in their buggies. What

You may wish to join a mounted drill team, which can be a fun way to meet other horse people and will allow you the chance to show off your horse at special events.

Halter classes offer a fun competition in their own right, but can also be a great method of introducing a young horse to the abundant stimuli of the show grounds without the pressure of trying to perform a more difficult job. The judge will want to observe the movement of each horse at a trot, so a segment of the halter class will be devoted to this. Be sure to practice beforehand with your horse, as it can take some careful work to get him to trot easily in-hand.

A beautiful team of horses hitched as a tandem are competing in a carriage driving class. Driving is a huge favorite of many competitors who enjoy the challenge of handling single or multiple horses or ponies from the driver's seat.

A beautiful sunset plays out behind the shadows of a horse and rider completing a ride at the end of the day. It's often the beauty of the horses and the environment where you spend your time that draws so many people to this lifestyle.

Tack stores often bring a selection of items to horse shows to appeal to an interested audience. Many equine enthusiasts enjoy taking a few moments from the fun of the show to do a little shopping.

To many horse owners, trail riding is the ideal way to enjoy a horse. There's nothing like wandering carefree throughout the woods or mountain trail. If you are interested in trail riding, then purchasing a horse who has already been introduced to this discipline is a good idea since you will require a sure-footed animal who is not easily spooked.

it means is that interest in equestrian driving is increasing all the time, whether it's enjoying a drive on the farm or participating in a driving show against stiff competition.

Driving can be, at its simplest, a horse hitched to a small cart. At its most impressive, it can be a four- or six-horse hitch pulling an enormous carriage or wagon. Other configurations are pairs (two horses, hitched side by side), tandems (two horses, one hitched in front of the other), and unicorns (three horses, with one in front and a pair behind). There are drivers who specialize in carriage driving, others who prefer pleasure driving, some who like

draft driving, and drivers who specialize in fine harness or roadster driving.

The American Driving Society (ADS) was founded for the promotion of carriage driving. The organization's website states that there are over 100 driving clubs across the United States, with over 60 of these being ADS-affiliated clubs. These clubs host driving events and competitions and bring together enthusiasts who share a passion for driving horses.

Trail riding with others is not only more fun, it's safer for everyone. Most horses enjoy the company of other equines and will more happily follow the trail if they have a friend to go with them.

Also gaining in popularity are Combined Driving Events (CDEs), which are three-day competitions, similar to eventing, where the competition consists of driven dressage, marathon (cross-country), and obstacles (cones). These competitions are truly the test of a versatile driving animal and are fantastic events for exhibitors to enjoy.

CAMPING AND TRAIL RIDING

If you are fond of camping, it's natural that you might consider the idea of taking your horse along on your wilderness excursions. Many people enjoy camping with their horses, trail riding through the woods, and experiencing the outdoors together. If your horse has a steady temperament and is good on trail rides near your home, he might be a good candidate for taking along on a camping trip.

You will want to make sure that you're fully prepared before embarking on such a trip. Much like preparing for a show, you'll want to be sure that you have everything with you that you might need. Unlike a show, you won't be near any stores or other horse people in the event that you forget some-

thing vital. Make your list and check it twice because forgetting something important can really put a damper on your trip!

Safety is, of course, the most important thing for yourself and your horse. Make sure that he is securely tied at all times, introduce him ahead of time to any unusual sights and sounds that he might encounter on the trip, and be certain that he is up-to-date on all of his vaccinations and dewormings before you go. Always take along his negative Coggins paper, as well as an interstate health certificate, if you are traveling out of state.

Bring along more than enough hay and grain, and it's also wise to haul water from home so that your horse has the option of drinking the water that he is used to. If he's at all reluctant to drink the water that is available out on the trail, such as in a stream, you might put some in a bucket and mix it with the water that you've brought from home to make the taste more familiar.

If camping overnight isn't quite your cup of tea, you can still experience the great outdoors on a smaller basis by going for a trail ride. Trail riding can be a relaxing, enjoy-

Dressage is the foundation upon which all other riding activities are based, Western or English. Here a top-level horse and rider demonstrate a difficult maneuver during a complex test. Dressage tests range widely from novice to Olympic-level skills to give everyone a chance to perform at a difficulty suitable for their abilities.

able way to spend the day, and best of all, you're back in your own bed at the end of the day!

The number one rule of trail riding (or camping) is that you should never go alone. You need another person along for safety reasons so don't go without a friend. Besides, all of those hours walking along the wooded path will be much more interesting if you have someone to talk to. Horses are great at listening and great at secret keeping, but they are quite poor at keeping up their end of a conversation.

In addition to your friend, make sure that you bring along a cell phone. You don't want to be caught in the wilderness somewhere without the ability to call for help if necessary. Bring along snacks and drinks for yourself and

the tools to remove a horse shoe if necessary. Miles into the forest is not a great time to realize that your horse has lost a shoe, but the situation will be considerably less stressful if you have the tools needed to rectify it.

Be sure to bring along plenty of first-aid equipment. In the event that your horse encounters any sort of scrape or cut while out on the trail, you'll want to be able to fix it up immediately. Also stick a roll of duct tape in your saddle bag. You will be amazed at the multitude of uses you will find for duct tape.

DRESSAGE

For many people, the word "dressage" conjures up images of the Olympic Games with fabulous riders and phenomenal Warmbloods completing their dressage tests in the magical Olympic setting. While this is undeniably one of the pinnacles of dressage competition, it does not mean that the discipline is inaccessible to the average rider with an average horse. Dressage, in French, means training. Dressage is the foundation for all other types of riding because it emphasizes the fundamental basics of balance and suppleness that are necessary for all types of equestrian sports.

The United States Dressage Federation (USDF) boasts over 30,000 members, which illustrates the great interest and support that dressage has among equestrians. This is due in part to the fact that dressage training can benefit any horse at any stage of training, from a green horse all the way up to a Grand Prix dressage champion. There are many competitions across the country where you can compete in dressage tests, from simple schooling shows all the way up to major dressage events.

A well-trained, upper-level dressage horse is a beautiful sight to behold. The intense level of communication between horse and rider with the subtle, scarcely noticeable cues that prompt amazing displays of athleticism and talent. These are the reasons that dressage has long been a perennial favorite in the world of equestrian sports.

THE HORSE TRAILER

If you're going to take part in all of these fun activities, you're going to need a way to transport your horse. You can look into borrowing a trailer from a friend or renting

The two-horse side load. A common trailer is very suitable for most horse owners who only own one or two animals. When buying a trailer, always look for rubber safety guards on key areas, such as near the doorway where the horse will have to step up. Trailers with a ramp on the back for horses to walk up instead of stepping up are also popular.

one, but the convenience of owning your own trailer is an asset that many horse owners feel is a "must-have." Even if showing is not your cup of tea, there will still be instances when you want to take your horse somewhere (to the national forest for a trail ride, to the vet for vaccinations, or to the breeding farm to be bred), and you will be glad to have your own trailer available to use whenever you want it.

Your choice of trailer will depend on what type of vehicle you will use to tow the trailer. Horse trailers come in two varieties: goose-neck and bumper hitch. Your vehicle's hitch will determine which type of trailer it can pull. Goose-neck trailers can only be towed with a pickup truck, whereas a bumper hitch can be pulled by a variety of tow vehicles. There are additional options such as aluminum or steel, straight load or slant load, and step-up or ramp. You will have to decide whether you want to spring for the extra amenities of a tack room and/or living quarters or if a simpler model will suffice. If you have only one horse, there's certainly no need for you to run out and purchase a four-horse trailer when a smaller model will be sufficient for your needs.

If you decide to purchase a brand-new trailer, you will want to get a good idea of the options available before making your choice. Spend some time browsing the

Small trailers can often be pulled by a vehicle you may already own, which can save money and maintain easy maneuvering. Be sure your vehicle is rated to tow the load you will be carrying.

Leadline classes are offered at most open shows and give a chance for very young riders to experience the excitement of competing and winning ribbons and prizes. The child retains some control over the animal while being lead by a person on the ground for safety.

websites of trailer manufacturers or visit trailer dealers and examine several different models so that you can judiciously choose the one that best suits your needs.

Used trailers are widely available. Horse owners are constantly upgrading their equipment, so there are many options to choose from if you are considering a used trailer. Ask around at the local riding and boarding stables or watch the ads in the newspaper. Some dealers who sell new trailers also deal with used and you may be able to get a good bargain on a lightly used model. Be sure to thoroughly evaluate the condition of any used trailer that you purchase, including checking underneath the mats to examine the floorboards, looking for rust, and checking the condition of the tires.

Once you've purchased a trailer, you will need to perform regular maintenance to ensure that the trailer stays in good condition and is safe for travel and for your horses. You will want to regularly inspect the hitch, check the air in your tires, and inspect the floorboards. It's wise to check your brake and signal lights before each use. A few moments to double check everything will save you trouble later on. It is best to have a place to keep the trailer indoors. Protection from the elements will certainly keep your trailer in better condition than if it is left outside in the rain, sleet, and hot sun.

Many consider the training and conditioning of their horse and themselves to be the biggest reward, but no one would turn down such an attractive prize as this!

RESOURCES AND BOOKS

Equine Color Genetics by D. Philip Sponenberg (Blackwell Publishing, 2003)
—A thorough study of the genetics behind equine color, this is an excellent book for breeders who are interested in learning the science behind their horse's colors. Contains a section of color photographs.

Feeding Horses and Ponies by Susan McBane (David-Charles Publishers, 2000)
—A thorough guide to feeding problems, types of feed, and balanced diets. A good reference volume.

Horse Color Explained by Jeanette Gower (Trafalgar Square Publishing, 2000)
—A delightful, full-color book with great information for anyone interested in learning more about horse colors.

Horse Owner's Field Guide to Toxic Plants by Sandra M. Burger (Breakthrough Publications, 1996)
—Fully illustrated with color photographs, this book will help you identify whether or not a toxic plant is invading your horse's pasture. Another excellent volume for your reference library.

Horse Owner's Veterinary Handbook by James M. Giffin, MD, and Tom Gore, DVM (Howell Book House, 1997)
—This is a great book that belongs on the shelves of every horse owner's library. It is a completely thorough and informative volume packed with information on all aspects of veterinary care for horses, from emergencies and diseases to foaling, parasites, and much more.

The USPC Guide to Conformation, Movement and Soundness by Susan E. Harris (Howell Book House, 1997)
—An excellent and informative book that covers the components of good conformation, highlighting conformational flaws, and understanding movement. Filled with wonderful illustrations, this is an excellent resource.

MAGAZINES

Equus Magazine
—A subscription to this magazine will keep you up-to-date on the latest in equine health and care. Filled with valuable information, this is an excellent publication to consider.

The Horse Magazine
—This is an informative publication with an emphasis on equine health. Full of good articles and information, this is a worthwhile publication. Website address: www.thehorse.com

WEBSITES

www.americandrivingsociety.org

—If you're interested in learning about driving, then the American Driving Society website is the place to start. Information on local clubs, membership information, and plenty of good advice to get you started.

www.completefoalingmanual.com

—This website featuring the *Complete Foaling Manual* contains an amazing archive of advice columns filled with questions and answers regarding pregnancy, foaling, and raising a foal.

www.doubledilute.com and www.equinecolor.com

—Excellent websites with information for color enthusiasts.

www.equine-reproduction.com

—This website contains a lot of informative articles on all phases of equine reproduction, including the breeding of your mare and artificial insemination. A wonderful resource.

www.equisearch.com

—This website puts a vast collection of information about horses at your fingertips. There are many fascinating articles and a lot of answers to questions that you may have.

www.foalnet.com/calc.html

—A handy calculator allows you to determine your mare's due date by entering her breeding dates. There are also calculators on this site that can quickly determine the number of days she is currently in foal or approximate dates for ultrasounds and vaccinations.

www.MDbarns.com

—A source for steel or metal barns, this website is a good place to begin looking for information on barns.

www.miraclewish.com

—Quality Welsh Ponies and Cobs, including well-bred young stock and trained children's ponies.

www.mortonbuildings.com

—Morton Buildings is a great place to start if you're interested in putting up a steel or metal barn, a pole building, or a storage shed.

www.usdf.org

—The official website for the United States Dressage Federation. If dressage is your cup of tea, this is the website for you.

www.usef.org

—If you're interested in competitive showing, you will probably want to become a member of the United States Equestrian Federation. Even if you're just browsing for information, you're sure to find plenty of it on this vast and informative site.

www.wdstar.com

—Woodstar Products, manufacturers of stall components and accessories, have a great website to visit if you're thinking about building or renovating a barn.

INDEX

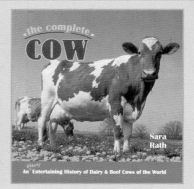

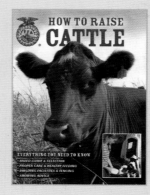
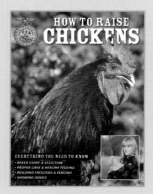
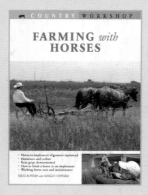